EARTH SHELTERED HOUSING

Principles in Practice

EARTH
SHELTERED
HOUSING Principles in Practice

MAX R. TERMAN

Illustrations by Virleen Bailey

VAN NOSTRAND REINHOLD
———— NEW YORK

Library of Congress Catalog Card Number 85-3189

ISBN 0-442-28287-7 (cloth)
ISBN 0-442-28288-5 (paper)

Printed in the United States of America
Designed by Ernie Haim

Van Nostrand Reinhold
115 Fifth Avenue
New York, New York 10003

Chapman and Hall
2-6 Boundary Row
London, SE1 8HN, England

Thomas Nelson Australia
102 Dodds Street
South Melbourne 3205
Victoria, Australia

Nelson Canada
1120 Birchmount Road
Scarborough, Ontario M1K 5G4, Canada

16 15 14 13 12 11 10 9 8 7 6 5 4 3 2

Library of Congress Cataloging in Publication Data

Terman, Max R., 1945–
Earth sheltered housing.

Bibliography: p. 194
Includes index.
1. Earth sheltered houses. I. Title.
TH4819.E27T47 1985 690′.8 85-3189
ISBN 0-442-28287-7
ISBN 0-442-28288-5 (pbk.)

To my wife, Janet,
   and daughters, Katy and Kerry

—*may flowers, trees, birds, and other members*
   of God's creation ever grace their world.

CONTENTS

PREFACE

One of our most important objectives as humans is to discover and pass on ways of living with our environment. Every form of life, including human, depends on nature's ability to produce clean air, pure water and fertile soil and to recycle wastes. It is our duty to live in a manner that enhances and preserves these natural processes. Earth sheltering—the use of earth cover to moderate and improve living conditions in buildings—is an old but recently rediscovered technique. It holds much promise for allowing us to use less energy and preserve more space for natural and human needs. It also gives the individual and society alike a real way to achieve self-reliance and independence from limited sources of fossil fuels.

An immediate need exists for information from a scientific yet practical perspective to fill the technical and philosophical gap between the scientific literature of earth sheltering and the popular writings of magazines and newspapers. So much of the scientific literature is theoretical and untried while the popular press has tended to be overly optimistic and uncritical. Basic questions need to be asked and actual experiences analyzed in order to arrive at some useful recommendations.

In this book I have attempted to answer questions that architects, engineers, builders, contractors, and potential owners would ask. I have provided basic principles from scientific journals and books and summarized the experiences of people actually living in earth shelters.

In the growth and development of any field of knowledge, there comes a time when theory requires testing, when concepts need to be refined in the light of experience. Such is the case with earth sheltering. A good foundation of principles has been laid, mostly through the excellent efforts of such organizations as the Underground Space Center of the University of Minnesota, the School of Architecture at Oklahoma State University, and the many design professionals active in earth sheltering. These principles are being put into practice by a small group of homeowners who are coming to understand both the positive and negative aspects of living in an earth sheltered home. We learn from their experiences, and because I have gone through this process, I also include information from my own experience to which many potential owners should be able to relate.

Most of the book is focused on the standard, elevational-type, earth-covered house, which has proved to be the most popular and energy efficient for cold and temperate climates. However, other designs are considered and the pros and cons analyzed. The appendices supply information on the actual performance of my own earth sheltered home, a list of professionals involved in earth sheltering, examples of award-winning designs, and a valuable accounting of actual and expected costs.

This book is not a step-by-step recipe for building an earth shelter but rather a source of information that will help interested people to avoid making major mistakes. The need for professional involvement is assumed, and information for choosing personnel is frequently provided.

I hope that this book will be an impetus for producing more principles and experiences in earth sheltering. Only when people involved in earth sheltering record their ideas and experiences will the store of knowledge grow. I have encouraged an ecological and ethical perspective for earth sheltered research. The problems that we face in energy and environment are crises in perspective as much as they are of technology. Earth sheltering, as a passive and environmentally benign technology, visibly expresses this perspective. The home is the central focus of our lives. It is our single most expensive purchase and the one possession that states our views about ourselves and our world. So, although this book is about earth shelters, it is about more than just houses.

Many thanks go to Tabor College, a singular institution with vision enough to provide its faculty with the time and resources to try to expand their horizons. I wish to acknowledge the help of Lon B. Simmons, an enthusiastic and knowledgeable professional in the field who not only has a feel for earth sheltering but for the underlying ecological and environmental realities. Simmons is credited with pioneering the use of post-tensioning in residential earth shelters and introduced me to the principle. He was invaluable in transforming my ideas into actual blueprints, construction details, and eventually, the reality of our home. Thanks also go to Terry Clark of Traverse City, Michigan who shared many ideas with me and to John Hatzung of Berg and Associates of Plymouth, Minnesota, who provided drawings and sketches of some of their work. I also wish to thank the many builders, owners, and researchers of earth sheltered houses who so freely shared their knowledge and ideas with me. Special thanks go to Virleen Bailey, who selflessly devoted many hours to preparing the illustrations and editing the manuscript. I also wish to thank the editorial staff of Van Nostrand Reinhold, whose concern and care were evident throughout the preparation of this book.

CHAPTER 1

ENERGY, ECOLOGY, AND HOUSING

And your ancient ruins shall be rebuilt; you shall rise up the foundations of many generations; you shall be called the repairer of the
breach, the restorer of streets to dwell in.   *Isaiah 58:12*

A house is not a home unless it accomplishes two tasks: providing shelter for its occupants, and preserving the environment of which it is a part. Modern houses appear to do only the first and to fail miserably at the second. In fact, the typical American house is becoming part of the problem rather than the solution, both economically and ecologically.

The flick of a switch or turn of a thermostat dial will no longer solve our home heating or cooling problems. With fossil fuels such as oil, gas, and propane becoming scarce and expensive, new ways that are less reliant on familiar energy supplies and more in tune with natural, sustainable sources of energy will have to be discovered. Passive solar and earth sheltered homes rely on design features rather than mechanical systems for heating and cooling and, therefore, make significant contributions to energy and environmental solutions.

THE ENERGY SITUATION

Any discussion of energy includes several terms that are basic to an understanding of energy relationships. For example, the term energy means the capacity to do work. In houses this is most often seen as a change in temperature. The units of energy most often used are British thermal units (Btu)—a measure of heat energy—and Kilowatt hours (kwh)—a measure of electrical energy. The two units can be used interchangeably with other units of energy as table 1-1 illustrates. A Btu in practical terms is about equal to the heat given off by burning one kitchen match. This amount of heat will raise the temperature of one pound of water 1° F. One kwh equals 1000 watts used for one hour and is equivalent to 3,412 Btu.

TABLE 1-1
Energy Terms, Conversions, and Content of Fuels

Conversions				
	kwh	**joules**	**calories**	**Btu**
1 kwh	1	3.6×10^6	$.86 \times 10^6$	3,412
1 joule	$.278 \times 10^{-6}$	1	.239	.000948
1 calorie	1.16×10^{-6}	4.18	1	.00397
1 Btu	.000293	1054	252	1

Energy Content of Fuels				
1 metric ton of anthracite coal	7,630	—	—	26×10^6
1 barrel of crude petroleum	1,641	—	—	5.6×10^6
1 pound of wood	—	—	—	7,140
1 pound of natural gas	—	—	—	26,000
1 barrel of gasoline	—	—	—	5.25 million

Solar energy can also be expressed in Btu. For example, about 950 Btu fall on one square foot of roof per day in January in Kansas. When this energy is converted directly into space heat (radiant energy into heat energy), the efficiency of the transfer is high. Only one transformation takes place. If the energy were in the form of coal (chemical energy, originally from the sun, processed by plants that fossilized) and changed into electric heat, the efficiency of transfer would be much lower because of the number of transformations (coal to heat to steam to electricity to space heat for the home). This is an example of how the laws of thermodynamics affect energy use. The fewer the steps between the source of energy and the end use,

the higher the efficiency will be. This is one good reason for using solar energy for space heat.

Figure 1-1 illustrates the basic concepts used in discussions of heat transfer and storage and building design. These processes are commonly observed in everyday living. An understanding of their application to passive and active solar design can be derived from figure 1-2. Passive systems by design have few moving parts and require little maintenance or external energy input such as fans and motors. In active systems, panels are added to a conventional house and depend on mechanical systems and controls to maintain temperature levels.

Far more solar energy falls on the roof of a

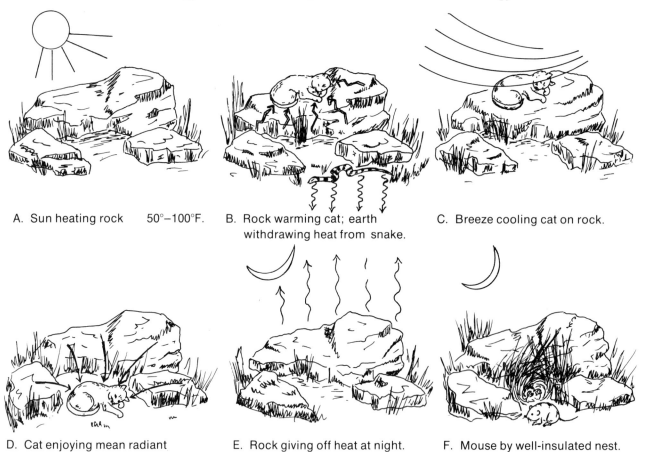

A. Sun heating rock   50°–100°F.

B. Rock warming cat; earth withdrawing heat from snake.

C. Breeze cooling cat on rock.

D. Cat enjoying mean radiant temperatures.

E. Rock giving off heat at night.

F. Mouse by well-insulated nest.

1-1. Principles of heat transfer and storage. A. Thermal radiation—transfer of heat between objects by electromagnetic radiation. B. Conduction—the transfer of heat between objects by direct contact. C. Natural convection—the movement of heat through the movement of air or water. D. Mean radiant temperatures—the average temperature experienced from the combined effect of all surface temperatures. E. Thermal mass—materials that can store heat and release it slowly. F. Insulation and R-value—materials that conduct heat poorly and reduce heat loss from an object or space. The R-value is a measure of the insulating ability—the higher the R-value the less the heat loss.

ACTIVE SOLAR DESIGN

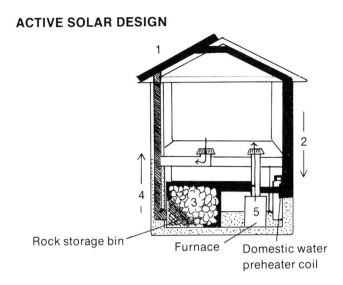

Rock storage bin
Furnace
Domestic water preheater coil

PASSIVE SOLAR DESIGN

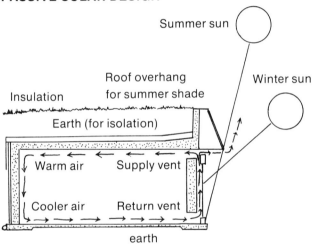

Summer sun

Roof overhang for summer shade
Winter sun
Insulation
Earth (for isolation)
Warm air   Supply vent
Cooler air   Return vent
earth

1-2. Principles of passive and active solar design. Active solar design: 1. Collector absorbs full sun. 2. Collector treated air flows to rock storage bin. 3. Rocks absorb heat; heat is retained in storage bin. 4. Cooler air in storage returns to collector. Passive solar design: 1. Vent with three modes: winter—closed with insulating panel; spring and fall—opened and closed as needed; summer—opened, supply and return vents closed. 2. Tempered-insulated glass fronting Trombe wall.

house in a single day than is needed to meet its heating needs for a year. The sun has always supplied the earth with a bountiful flow of radiant energy. Fossil fuel resources, however, are limited, and consumption can readily outstrip production (see fig. 1-3).

Today, oil provides over a third of the energy used in the world annually and about half

that used in the United States. (New York City, for example, depends on oil for 71 percent of its needs.) Until 1973 there appeared to be no problem. Then, in that year, the oil embargo ended the age of cheap, readily available fossil fuel, and oil prices have jumped from three dollars a barrel before the embargo to about twenty-nine dollars a barrel by 1983. The effect was most obvious at the gasoline pump, but the home was also affected as heating and air-conditioning costs soared.

The American energy system is huge and complex. In 1976, the United States burned 75 quadrillion Btu (quads). The oil used would have filled a half-million-acre lake to a depth of two feet. The heat from the energy used in one year theoretically could raise the temperature of the Pacific Ocean 11° F.

Thanks to energy conservation, there presently is an adequate supply of oil in the United States; some people even worry about an oil glut. However, there is no way of predicting the adequacy of supplies even five years into the future. Geological unknowns, political unpre-

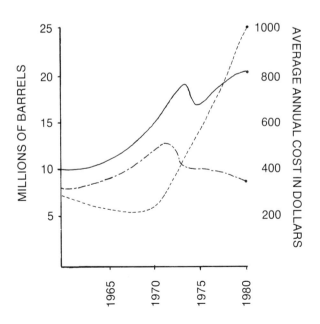

1-3. The average annual cost of heating a home with oil in the United States is represented by the dotted line, the daily domestic supply by the dash-dot line, and the daily domestic demand by the solid line.

dictability in the Middle East, and changing consumptive patterns have left the experts in disarray. Recent discoveries of crude oil in countries not belonging to the Organization of Petroleum Exporting Countries (OPEC) were fortunate but may not be repeated. The present concern is that abundant supplies now will cause the United States to forget 1973 and, thus, the future. As long as we rely on imported oil, we will never control our own energy destiny. Of the 670 billion barrels of proven oil reserves in the world, OPEC controls 450 billion of them.

Furthermore, other countries are demanding an increasing share of the fossil fuel supplies to support their hopes for a Western-style economy. The Third World countries are likely to double their demands by the end of the century, and 80 percent of the known supplies lie in these countries. Russia has 10 percent, and the United States and Europe combined have just 9 percent of global oil reserves. Obviously, reliance on this extremely concentrated and unpredictable resource is an invitation to disaster.

Natural gas, although largely untapped, is also an unequally distributed resource. Most gas resources are located in just four regions—Mexico, Russia, the Middle East, and North America. Most of the United State's reserves have been tapped, and production will likely decline in ten years. The same is true for Europe.

Natural gas is a concentrated resource that is not as easily shared as oil—liquefied gas is costly and dangerous to transport, and gas pipelines are limited by the geographical nature of the supplies.

Coal is relatively abundant and will probably overtake oil as the world's primary fuel by 1995. Its use is expected to double or triple in the next twenty years, mainly for producing electricity and steel. The most severe limitations to coal use are environmental and health related. Burning coal contributes significantly to air pollution and acid rain problems.

Conversion to synthetic fuels is very costly, and their use is inefficient. Coal and its derivatives will probably never replace oil com-

pletely, although methane and methanol may do so, especially in transportation.

Opposition to nuclear power, an energy alternative, has grown steadily since 1960, reaching a peak in the mid-seventies and eighties. The problems concern high costs, public safety, waste disposal, and weapons proliferation. Between 1971 and 1982 nuclear plant costs rose 142 percent, and between 1980 and 1983 electric rates increased 40 percent. Both are expected to increase another 20–60 percent when more of the nuclear plants presently under construction come on line. Because of the tremendous costs involved, nuclear power will probably only supply 3–5 percent of the world's power in the year 2000. A nuclear power plant starting construction in 1983 would not be completed until 1995, at a cost of approximately 10 billion dollars. The "too-cheap-to-meter" fuel of the 1950s has already become the "too-costly-to-use" power of the 1980s.

Table 1-2 gives the present usage and predicted depletion dates for our nonrenewable fuel supplies. Although these dates vary from one expert to another, there is general agreement that our fuel-powered age will not last beyond the year 2100 and will probably end well before then. For practical purposes, the age of oil will not end when the last drop of black liquid is pumped from the ground but when the price for oil gets so high that no one is able to afford it. In other words, the country will "run out" when it takes more energy to acquire the oil than can be derived from it.

TABLE 1-2
Usage and Predicted Depletion Dates of Fuels

	Percentage of Total Use	Depletion Date
Oil	46%	2015–30
Natural gas	24%	2025–60
Coal	20%	2100
Uranium	3%	2040

In the past, the market costs of energy did not reflect the real costs. The production of fossil fuels was subsidized by tax credits and allowances, and the price to the consumer did not reflect such costs as environmental pollution cleanup charges. Oil once returned ten times the energy used to produce it. This 10-to-1 ratio is now approaching a 5-to-1 ratio and will keep shrinking as oil supplies are depleted.

Materials, too, are affected by the diminishing fuel base and escalating prices (for example, aluminum with its energy-intensive processing requirements). As energy gets higher in price, so do the vital materials of our consumer goods. Energy and materials operate together to affect standards of living.

Official response to the shortages has been to expand the sources of energy (oil, coal, natural gas, and nuclear) to meet the peak demands of present "energy sieves." Utility companies maintain that with more plants and fewer environmental restrictions they can meet all consumer demands. Consequently, they have invested heavily in new exploration equipment, supertankers, ports, storage tanks, refineries, pipelines, and nuclear power plants.

Another hidden cost of energy is the incalculable expense to the environment and to human lives. An article in Science News estimated that for each 2 trillion kwh of electricity produced in 1976, 6,000 people died of air pollution, 1,250 died in associated occupational accidents, 10,000 to 1 million children developed respiratory disease, 60,000 to 6 million adults developed chronic respiratory disease, and 100,000 to 10 million people had pollution-related asthma attacks. Also, 100,000 to 200,000 acres of land needed to be reclaimed, and 1.6 million tons of carbon dioxide were added to the atmosphere. In the nuclear industry, 20 to 200 deaths were job related, 4,000 to 7,000 nonfatal illnesses developed, and 20,000 acres of land were scarred in mining. An additional 2,000 acres were used for waste storage.

Between 1977 and 1980, Americans did reduce oil imports 28 percent by using energy more efficiently and wisely. The United States consumed less oil in 1982 than in any year

since 1970; in 1983 it imported 27 percent less oil than in 1973 and bought more oil from Mexico, Canada, and Great Britain than from OPEC. As a result of conservation, the energy required to produce one dollar of gross national product has dropped by 28 percent since 1973. Oil once provided 15 percent of America's electricity—now the figure is only 6 percent.

Some experts estimate that Americans still waste between 80 and 90 percent of the energy available. For example, just by using more efficient light bulbs, Americans can save about 500,000 barrels of oil a day. This becomes even more important when considering that each American paid OPEC nearly $360 in 1980 for imported oil, a situation that severely affected the economic well-being of the United States.

American architecture is another energy drain. American buildings use 50–90 percent more fossil fuel energy than is necessary; 33 percent of the energy supply is used to operate a building and 15 percent to construct and maintain it. Forty-four percent of all energy used is associated with architecture. It could be put to better use. Further complicating this situation, Western countries have exported their energy-intensive architecture. In the past, many well-intentioned efforts have proved to be embarrassingly unsuccessful because the heating and cooling systems of Western-style buildings were complex mechanically and depended on the availability of an unfailing energy supply, available spare parts, and knowledgeable technicians. These things are rarely available in developing countries.

Our high energy consumption has also led us to affect the lives of those in foreign countries in other ways. Americans are accustomed to considering the energy problem as something that stops at the boundaries of the United States. A "fortress America" outlook has developed that focuses on our needs exclusive of those of other countries. In 1973 the average American used 357 million Btu of energy whereas the average person in the remainder of the world used 39 million. In 1980 the mean energy consumption per person in the United States was approximately fifty-four times that of the less developed nations, eight times the world average, and two to three times that of other industrial nations such as Japan and Great Britain.

In 1981 Americans used more energy for air conditioning alone than 985 million Chinese used for all purposes. Overall, Americans were responsible for 20 percent of the world's coal consumption, 49 percent of the world's natural gas consumption, and 25 percent of the world's total petroleum consumption. Yet we have only 5 percent of the world's population!

Beginning now, energy and mineral resources must be used as carefully as individuals use their personal finances. Energy's income resources are the various forms of solar energy, such as direct heat or light, as well as wood, wind, and so forth. Because these resources will last as long as the sun, they can provide a steady income on which to draw at the beginning of each new day.

Fossil fuels can be regarded as savings, accumulated over millions of years from deposits of solar energy. Since these stocks are finite, they should be used sparingly and only when needed for "investments" in solar technologies that will yield a return in increased income. The various mineral and geothermal resources (including nuclear fuels) should be considered an inheritance since they are a part of the planet and were so from the beginning. This legacy should be used and recycled in cooperation with the solar "income" to establish better ways to live sustainably and in balance with the environment. All of nature operates this way—living within its income of available food and sunlight while the mineral inheritance is invested and recycled.

THE ENVIRONMENTAL SITUATION

The great paradox of the modern age is the separation that has occurred between ourselves and the natural processes on which human survival depends. Quite literally, survival depends on developing harmonious rather than disrup-

tive forms of technology. The laws of nature have not been repealed. The challenge is to understand them in their complexity and adapt accordingly. This does not mean that all the materials and processes on which our society depends are abandoned. It does mean that the protection of life-supporting natural processes is primary.

According to the *Global 2000 Report to the President*, made in 1980 by the Council on Environmental Quality, the main environmental problems are soil erosion and soil deterioration, water pollution, forest losses, air pollution, rising carbon dioxide levels in the earth's atmosphere, disruption of the ozone layer in the stratosphere, radioactive contamination of the environment, and the loss of genetic diversity through species extinction and habitat destruction. In practical terms our country is losing farmland, forests, lakes, streams, wildlife, and those basic requirements of life—clean air and water.

In some places, the predominantly roofed and paved areas of our civilization have changed whole climates. In New York City, for instance, a coastal city, temperatures are higher, visibility less, solar radiation less, air more stagnant, and humidity higher in summer and lower in winter than in underdeveloped oceanfront regions. This is in addition to a tremendous loss of natural habitat. Earth sheltered techniques are especially relevant in urban situations because they make it possible for communities to construct buildings and still preserve green space to help modify the negative impact of the built environment.

The earth's surface, like other natural resources, is finite. We must be conservative in our use of it. Subsurface space is a vast resource to be explored and used. It already provides for transportation systems; power, water, and sewage lines; and gas and oil storage. When this option is used wisely, it may be possible to provide space for both humans and nature.

To grasp the magnitude of environmental problems, an understanding of basic ecosystem structure and function is necessary. Figure 1-4 is a diagram of a natural ecosystem incorporat-

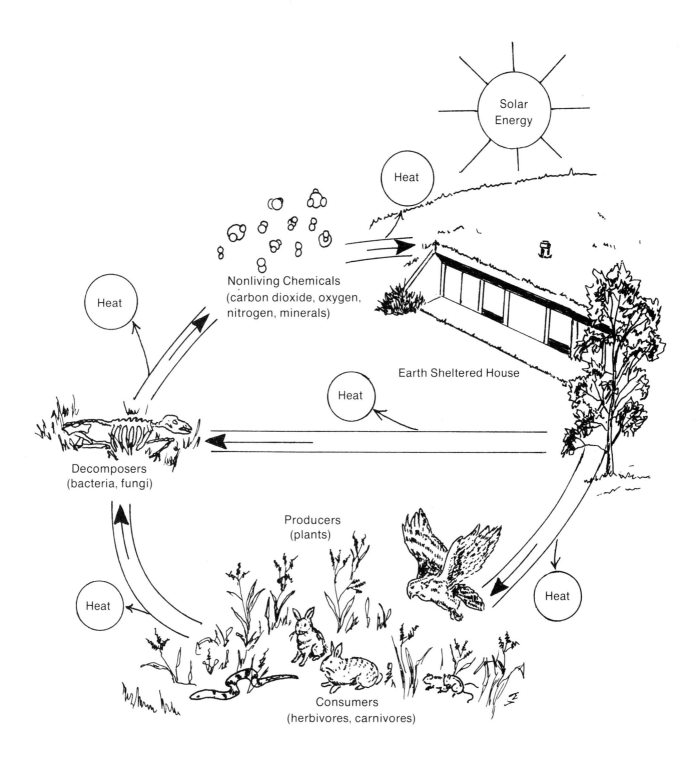

1-4. A summary of ecosystem structure and function incorporating a passive solar, earth sheltered house.

ing a passive solar, earth sheltered house. An ecosystem is a community of living things interacting with each other and with the physical environment (sun, water, air, soil, and chemicals). The green plants capture the sun's energy and turn it into food, such as sugar, and more plant material. Plant-eating animals, herbivores, eat the vegetation and use the energy to produce more animal tissue. Carnivores prey on the herbivores and convert the energy derived into their own flesh. As each of these living things excretes and dies, their wastes and bodies are broken down by bacteria and other decomposers. The elements in the bodies of animals and plants return to the soil where the green plants use them, thus closing the cycle. Energy from the sun drives the cycle, and the chemical elements flow from link to link in the food chain. As long as this chain remains intact, the ecosystem is balanced and stable, providing for the needs of each of its inhabitants, including the people who live in the house.

Until recently, everything man did to damage the environment was reversible because the human population was small and had little access to concentrated forms of energy. Now the human impact is global, and the demands placed on the environment are often beyond its regenerative capacities. Urbanization and exponential growth in population and energy use have meant that land, water, and air stay polluted, adversely affecting human health.

The National Wildlife Federation claims that the overall environmental quality is declining. By the year 2000, it is predicted that we will lose 20–40 percent of the world's forests and 500,000 species of plants and animals. Between 1967 and 1975, the United States lost over 6 million acres of cropland to urban and other "builtup" uses. These are the unpaid costs of our lifestyle that we are beginning to pay—not with money but with the declining quality of our lives.

Why are balanced ecosystems essential? Ecologically and aesthetically, there is a need for varied landscapes made up of crops, forests, grasslands, marshes, swamps, estuaries, lakes, streams, ponds, and other natural areas. Besides providing food and fiber, these areas perform necessary services. The oceans, forests, and grasslands act as a climatic buffer, purifying air and water. Vegetation also prevents soil erosion.

The ecosystem's basic functions of production, consumption, and decomposition provide for the soil's fertility, which is the basis of agriculture. Even the crop plants depend on wild ancestors for continued genetic viability through hybridization techniques. Most of the habitats supporting these wild plants are now rapidly being destroyed. According to E. P. Odum, a University of Georgia ecologist who has been seeking the proper "developed-to-natural" ratio for land use, at least a ratio of three to one between natural and manmade environments is required to maintain vital life support systems. In addition to these practical necessities, natural areas also provide space for recreation and wilderness experiences, which renew body and spirit.

Instead of perpetuating the conventional, supply-oriented solution to the energy and environmental dilemma, the new philosophy approaches the problem from an ecological perspective, relating energy use to the life it supports. This philosophy embraces an ecological approach to architecture which

— emphasizes the maintenance of ecological and cultural diversity and stability

— is simple and understandable

— provides creative roles for all sectors of society in the adaptive response to energy and environment

— invests capital, energy, and materials to build durable, efficient, and low-impact structures

— emphasizes self-reliant use of local resources

— adapts to local environment and culture

— emphasizes recycling and reclamation

— uses renewable energy sources (sun, wind, water, and vegetation) and is efficient in its use of nonrenewable resources (petroleum, coal, natural gas)

— emphasizes growth and creativity in areas such as research and education, maintenance and service, and production of durable goods

— has a global outlook and considers the needs of all humankind

This side postulates not an energy crisis but a demand crisis and advocates a decentralized policy of energy use and conservation that will not pollute; will not produce radiation, hot wastes, and atmospheric changes; will not deplete limited fuel supplies. There will be less strip mining, oil spills, explosions, and money paid to OPEC and more natural landscapes, clean air, clean water, and wildlife.

HOMES, ENERGY, AND THE ENVIRONMENT

Today's houses are designed as if they were independent of the environment. Mechanical systems are installed with the confidence that any problem can be met with a push-button solution. This design attitude has fostered outdated building codes, standards, money-lending practices, zoning ordinances, and utility companies. Furthermore, the outdated homes still being constructed today will burden the future with their high operating and maintenance costs.

A family living in a typical house often has to decide whether to turn up the gas furnace or pay the next mortgage installment. There is not enough money to do both because of the high cost of natural gas (see fig. 1-5). In 1970 the average energy bill for a family of four totaled about nine hundred dollars. This included energy for space and water heating, lights, appliances, and the family car. By 1980 this amount had climbed to three thousand dollars, with a projected average annual energy cost of five thousand dollars or more by 1985.

Not only do conventional houses use excessive amounts of fossil fuels that contribute

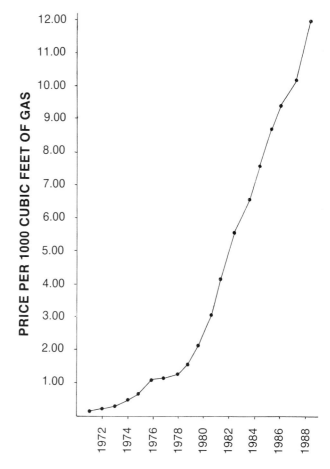

1-5. The escalating cost of natural gas.

to air and water pollution, but they are destructive to the land on which they are built. In figure 1-6, a typical housing development is under construction, clearly demonstrating the loss of wildlife habitat, groundcover and topsoil. The native plants are replaced by commercial varieties. In about fifty years, these houses will die, new ones will be built, and the damage begins anew.

Furthermore, about 17,213 square feet of wood are required to build a typical home. Lumber companies have found it difficult to keep up with demand. In 1977, the United States had to import over 22 percent of its lumber, and prices have been going up. Logging practices are contributing to soil erosion and loss of habitat, particularly in developing countries where wood is also used as fuel and deforestation is severe.

Figure 1-7 illustrates the necessary transition in housing. Solar-powered, earth sheltered homes have the potential to meet 90 percent of space-heating and cooling needs with renewable forms of energy. This is possible now. There is no longer any need to wait for new technology, although more research is necessary to make the technology generally available. According to the *Global 2000 Report*, there are from thirty to fifty years in which to engineer the necessary changes in energy systems to solar. It is important to act now while fossil fuels are still available to make this transition.

Buildings of the future, solar-powered age actually may not be too different from those of our early American ancestors. Virtually all vernacular buildings—those that are characteristic of each region—were buildings whose basic form allowed for natural heating and cooling by the environment. In essence, they were passive solar buildings.

In the Southwest the use of heavy masonry reduced the heat of the midday sun, storing and releasing it at night. Where the comfort conditions depended on capturing the wind, fewer walls were exposed to the sun and the prevailing wind was brought into each house by air scoops and strategic openings. In regions of cold, dark winters, windows and rooms became smaller, ceilings lower, and fireplaces massive. The summers in the South necessitated large, high-ceilinged rooms, wide porches, and overhangs—all of which allowed maximum shade and air circulation.

The basic reason for the wedding of form and function in the housing of the past lies in a realistic acceptance of and sensitivity to the environment. The modern gap between form and function in houses was created by cheap energy and the homogenization of culture. Dependent on fossil fuels, most houses are alike in style and vulnerability.

Spectacular innovations are not necessarily the best approaches to solving energy problems. T. S. Eliot's remark that "the end of all our exploring will be to arrive where we started and

1-6. Typical housing development.

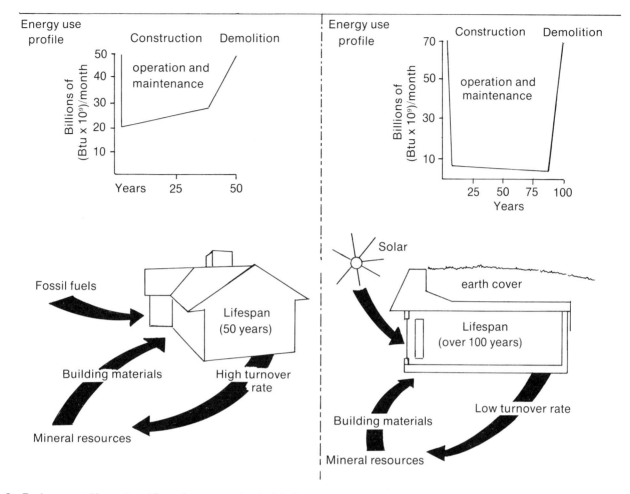

1-8. Environmental impact and flow of energy and materials for a conventional and a passive solar, earth sheltered house. Energy use is expressed in billions of Btu used per month.

can be halted. If this can be done, then the victories won in that war may seem small in comparison.

Firsthand Experiences

As an environmental scientist and professional ecologist, I have come to the conclusion that the only way to save our ecological systems is to preserve surface area and habitat. In every natural process, the first stage begins with a form of vegetation that is growing in a medium of soil or water. Without saving these essential elements—vegetation, soil, and water—we will not be able to stop the degradation that is going on around us.

Until recently, my lifestyle was contribut-

ing to a way of life that wasted more than 80 percent of the available energy. Although I hold a doctorate in environmental science and ecology and teach in these fields, I did not incorporate these concepts into my own life until I was challenged, by one of my students, to "walk the walk if I was going to talk the talk." I became serious about reducing my family's consumption of energy, beginning with our home, the focal point of our physical existence. It represented our biggest major investment, both financially and emotionally.

Prompted by the shortages, the high cost of fossil fuels, and a belief that excessive consumption on the part of an individual (and a nation) leads to many injustices, our plan was

DEVELOPMENT OF A NATURAL SYSTEM

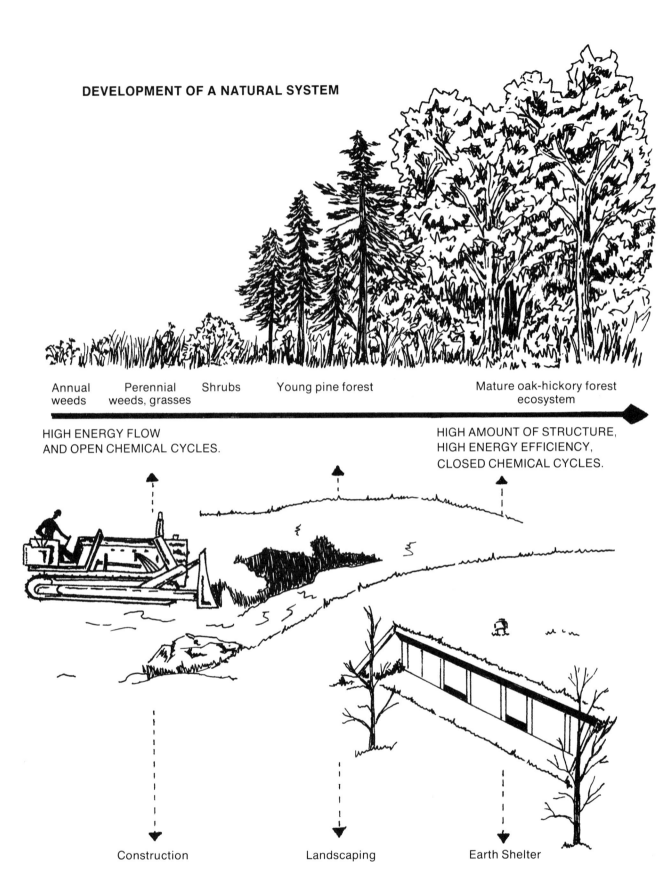

Annual weeds   Perennial weeds, grasses   Shrubs   Young pine forest   Mature oak-hickory forest ecosystem

HIGH ENERGY FLOW AND OPEN CHEMICAL CYCLES.

HIGH AMOUNT OF STRUCTURE, HIGH ENERGY EFFICIENCY, CLOSED CHEMICAL CYCLES.

Construction   Landscaping   Earth Shelter

1-9. The stages of development in a forest and in a passive solar earth sheltered home.

to wean ourselves from the fossil fuel economy as much as possible. A serious investigation of alternative housing options led to the conclusion that, in our climatic region, a passive solar, earth sheltered home would be best. This was an ethical as well as an economic and scientific decision to shift away from a materialistic and consumer-oriented value system in an effort to live peacefully and in balance with the rest of creation.

With this decision, my lifestyle is in line with my teaching, and both are more convincing because guidelines are now lived as well as taught.

Our home, whose design, construction, and performance will be detailed in later chapters, is set into a south-facing slope on the Kansas prairie. The surrounding vegetation is gradually returning to what it was before the excavation and construction. The water run-off patterns are not disturbed, and the native wildlife is slowly regaining former levels of abundance and distribution. The area is used as an environmental study area in my ecology classes.

From the air, the house is barely noticeable, a mere "smile" in the hillside. The climatic impact (reflective characteristics, known as albedo, and thermal and humidity influences) of this homestead is indeed gentle. Aldo Leopold, noted ecologist and land usage expert, has said, "A thing is right when it tends to preserve the integrity, stability, and beauty of the biotic community." He might have been speaking of the earth shelter.

CHAPTER 2

AN ADAPTIVE APPROACH TO HOUSING OPTIONS

The naive believes everything, but the prudent man considers his steps.
Proverbs 14:15

The selection of a suitable energy-efficient home to build is not easy to make. More than money and energy savings are involved. There are also reasons of aesthetics, land use and environment, noise reduction, maintenance, and protection. Furthermore, no single housing strategy or technique will be best in all situations. Site conditions and climate may favor one design feature over another.

Therefore, it may be necessary to select a combination of such strategies as conventional, active and passive solar, and earth contact. Some perform better in cooling (earth contact) whereas others rate extremely well in heat retention (superinsulated types)—in short some designs do well in some regions and not in others. There are no simplistic answers.

The ultimate goal for the home of the future should be attractiveness, energy efficiency (both in heating and cooling), and low maintenance—a structure that functions well with its environment. The home must be durable and must not denude the area of vegetation, cause undue water run-off, or destroy excessive amounts of wildlife habitat.

PRINCIPLES: THE HOUSING OPTIONS

By comparing aboveground and earth sheltered approaches to housing and discussing some of the pros and cons of the various forms of energy-efficient housing, it is hoped that the best attributes of all forms can be applied, once the benefits and limitations are appreciated. Earth sheltered construction, because of the energy and ecological advantages, is unique and thus the focus of the book.

Figures 2-1, 2-2, and 2-3 illustrate some possible options open to present and future homeowners. However, there are many variations on these themes, and as stated previously, no single form can solve all problems. Instead of being static and unchanging, the home becomes part of an adaptive, dynamic process. Climate-sensitive designs, tailored to specific regions, turn houses into organisms—changing and adapting, filling niches, and becoming parts of the ecosystem.

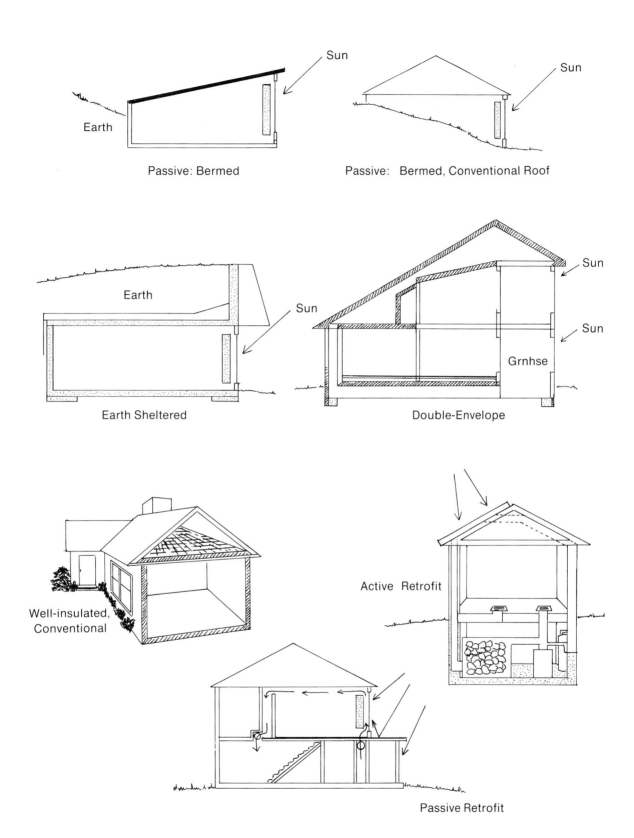

Sun

Earth

Passive: Bermed

Sun

Passive: Bermed, Conventional Roof

Earth

Sun

Earth Sheltered

Sun

Sun

Grnhse

Double-Envelope

Well-insulated,
Conventional

Active Retrofit

Passive Retrofit

2-1. Some housing options. Arrows indicate angle of sun rays
or heat flow.

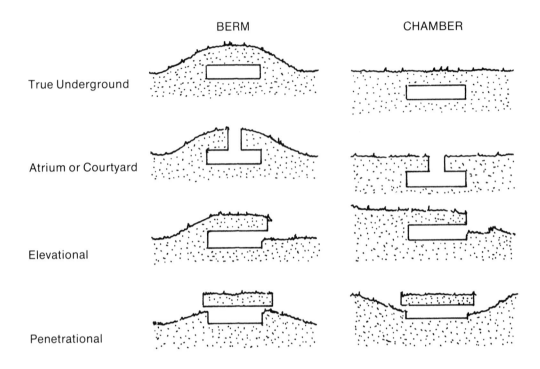

BERM   CHAMBER

True Underground

Atrium or Courtyard

Elevational

Penetrational

2-2. Types of earth covered structures. (Source: Ken Labs, 1975)

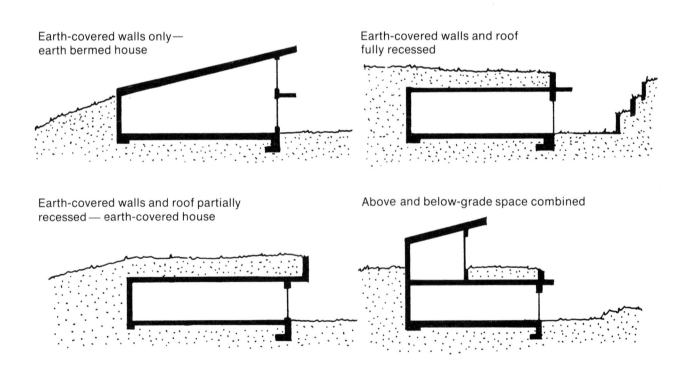

Earth-covered walls only—
earth bermed house

Earth-covered walls and roof
fully recessed

Earth-covered walls and roof partially
recessed — earth-covered house

Above and below-grade space combined

2-3. Some variations in earth sheltered structures.

Conventional Housing Strategies

For owners of conventional homes, table 2-1 lists the conservation and retrofit possibilities. Although these are certainly very effective (energy savings of up to 80 percent can be achieved), they do not meet the cooling, durability, maintenance, and ecological requirements of homes in the coming years.

For future homeowners, a wide choice of active and passive design strategies exists. These range from superinsulated, active or passive solar homes to double-envelope houses and earth sheltered structures. This chapter will evaluate the earth shelter as a housing strategy, but it should be remembered that many other types of energy-efficient homes are being built and are performing quite well. In reality, the earth sheltered concept is not a separate strategy from the others. It can often be combined with other designs.

Superinsulated Housing

A superinsulated house is one that is situated in a cold climate, receiving only a modest amount of solar energy, yet is so well insulated and airtight that it is kept warm by a combination of solar heat and other internal sources such as stoves, lights, and refrigerators. Superinsulated houses require less than 15 percent of the energy required by conventional houses. They can be built anywhere a typical house can be built and with a conventional appearance. Usually, they have less window area than the conventional home (south-facing glass is normally less than 8 percent of the floor area). With meticulous weatherstripping, every crack is sealed against infiltration, and a perfect vapor barrier is installed around outlets, pipes, and partitions. About two-thirds of the insulation is placed on the cold side of the wall and one-third on the warm side with the vapor barrier in between.

Superinsulated homes built of standard materials cost 5–10 percent more than conventional housing. The insulation and infiltration detail may require special engineering, and the building may be somewhat more difficult to

---

TABLE 2-1

Heating and Cooling Strategies for Conventional Homes
(Percent Energy Savings in Parentheses)

1. Weatherproofing

— insulate roof, walls, and floors (20%–50%)

— install double- or triple-glazed windows; insulative shades, storm doors and windows; and air-locked entries (10%–30%)

— weatherstrip all doors, windows, and air leaks (10%–30%)

2. Natural Heating and Cooling

— orient to receive sun, prevailing breezes; install operable windows, wind turbines, and vents

— plant deciduous trees and vines on south side and evergreens on north

3. Energy Adjustments and Modifications

— lower thermostat (each degree reduction can save 3% on heating bill)

— humidify in winter; dehumidify in summer

— install a woodstove (airtight) or fireplace insert

— insulate heating and cooling ducts; clean furnace filters

— install clock thermostats (for hot water heaters)

— use fans to move air; use exhaust fan to ventilate at night; close up house in the day

— install awnings and shades; install external shutters

— use light-colored roofing and siding to reflect solar radiation and reduce cooling load

— place air conditioners in cool, shaded areas; use only the brands most highly rated for efficiency

— use a heat exchanger in tightly sealed homes to exchange air without losing heat

— install a heat pump or high-efficiency furnace

— wear appropriate clothing; schedule activities to supplement heating or cooling loads (cook inside in the winter and outside in the summer)

— install active or passive solar retrofit systems to help heat the home after it has been well insulated and weatherproofed

---

construct than typical housing. The insulation keeps the home cool by retaining nighttime temperatures. However, summer overheating has been reported as a minor problem in some superinsulated homes. Humidity is another potential problem in both summer and winter because of the tight construction, and adequate natural lighting may require careful design to achieve.

Other potential disadvantages include: a limited outside view, a sense of confinement, reduced floor space because of increased wall thickness, and the need for heat exchangers (to exhaust stale air and to heat the incoming fresh air).

Double-Envelope Houses

A double-envelope house incorporates convective air circulation between inner and outer layers. In other words, as the name implies, there are two "envelopes" with a continuous air space of at least 6 inches between them. A greenhouse or sunspace is situated on the south side, and a large airspace below grade (crawlspace, basement, or subbasement space). Air circulates from the greenhouse, through the air passageway to the roof, and then down the north wall channel to the crawlspace and back to the greenhouse or sunspace. Ordinarily there are no ducts or blowers, although some studies indicate that the convection should be augmented by a fan.

Double-envelope homes appear to be in an early stage of development. There are still many questions mainly dealing with costs, heat storage, and distribution, and few detailed studies on actual houses to answer them. Excessive heat loss through the greenhouse and problems with ventilation and summer cooling have also been reported. Could the distribution and storage of heat be more effective and inexpensive in a well-insulated home using more direct, passive techniques? In a study reported by *Popular Science*, it was concluded that the main energy-saving aspects of a double-envelope house resulted from the insulative value of the double envelope and not air circulation. Hybrids of double-envelopes and superinsulated homes are now being developed that maximize the insulation and solar gain.

Active Solar Heating Systems

Active solar systems consist of panels that are added to an existing conventional or a new home. These panels circulate air or liquid in order to collect and transfer the heat, and are the first thing many people think of when they hear the term *solar*. Active solar panels have been in use for approximately thirty years. The collectors are usually for heating only and are often retrofitted to existing homes. Advances in coatings, glazings, and microprocessor controls have resulted in many systems that perform quite well. Active systems tend to be expensive, however, because panels must be added to an existing structure. The system is complex with many moving parts and the life span is not as long as that of the passive systems. Perhaps for these reasons, interest in active solar is waning, and people are now considering the advantages of the simpler and more reliable passive solar systems.

Passive Solar Houses

Passive features, such as direct solar gain through windows and sunspaces, are easily integrated with any new and some existing structures. A glazed exposure for collecting solar radiation plus a structure with substantial thermal mass for heat storage are required. The added glass and mass may increase the cost to slightly more than that of a conventional home, and special attention to the choice of a site is required since a solar exposure is needed.

Although heating is now the primary benefit of passive systems, means of cooling are being developed and will be discussed in later chapters. Potential disadvantages, when compared with conventional housing, include possible loss of privacy, higher visibility on site, greater requirement for control of heat loss, possibly greater temperature swings, and more difficulty in construction. However, the potential for a passive solar system is great since this strategy is so easily integrated with a number of other design concepts.

PRINCIPLES: EARTH SHELTERED HOUSING

Only one of the energy-efficient housing types in figures 2-1, 2-2, and 2-3 completely meets the heating, cooling, and ecological requirements—

the passive solar, earth sheltered structure. If properly designed, this type of home will meet nearly all its needs with natural energy and, at the same time, promote the stability of the ecosystem of which it is a functioning entity. Because of this dual benefit—energy and ecology—earth sheltering is the option highlighted in this book. Topics covered below are merely introduced at this point and will be detailed in later chapters.

Using earth to protect a house is far from a novel idea. In response to an arid, hot, and almost timberless environment, peoples in Turkey and Tunisia turned to earth shelters around A.D. 800. The Chinese have been living in earth shelters for at least five thousand years, and in the United States, the early American pioneers lived in sod houses and dugouts on the prairies in response to the severe climate and lack of building materials. In the Loire and Cher Valleys of France, several modern communities have been comfortably inhabiting caves for years.

Now that energy is expensive and land and timber are disappearing, the use of earth sheltering as a building strategy is again becoming popular. With modern construction methods and materials, the drawbacks of the sod houses and some modern basements (leaks, dampness, odors, and so forth) have been overcome, and there are presently more than three thousand earth shelters in the United States.

The environmental movement of the sixties and seventies joined with the energy crisis to spur a few architects to venture into designing and constructing modern earth shelters. Today, the list is long and growing (see Appendix B). The Underground Space Center of the University of Minnesota authored the first technical manual on earth sheltered design in 1978, and it soon sold over two hundred thousand copies. The Center is a national and international resource for the research, coordination, and transfer of information concerning underground space use. The center has received grants from such agencies as the United States Department of Energy and the Department of Housing and Urban Development to explore the relationship of earth sheltering to design and energy use, the

psychological effects of underground living, and policy and financial considerations. The Underground Space Center solidly supports the efforts to fulfill the promise of underground space utilization, an inherently simple concept that makes sense to a growing number of people.

The term *earth shelter* can have many different meanings. It has been used to describe: conventional homes with basements or split levels, earth-bermed structures, and totally underground buildings. For this reason earth sheltering is best defined functionally. The Underground Space Center states that an earth sheltered building is a structure that uses the earth in a deliberate attempt to benefit the environment of a habitable space. These benefits may be ecological, aesthetic, economic, and related to land use.

Earth sheltered structures may have different relationships to the ground surface. In some, only the walls are covered with earth whereas others have soil on the roof. Those with earth-covered walls but conventional roofs are earth-bermed structures; houses with a majority of the roof covered with soil are called earth-covered structures. Earth shelter is used to refer to both, but generally the term applies to earth-covered structures. Some earth sheltered homes are combinations of aboveground and below-grade structures. Others may have soil on the roof but little earth against the walls. Many have only one wall exposed, and some have a central space or atrium around which rooms are clustered. Figure 2-3 illustrates the relationship of various types of earth shelters to the ground surface. No doubt future innovative architects and designers will come up with many more variations.

Earth-Covered versus Earth-Bermed

The earth-bermed home represents a popular compromise between the earth-covered (soil on the roof) and the surface home. The earth-bermed house with a conventional roof does not have the same thermal and ecological potential of the earth-covered home. The soil and vegetative cover on the roof isolates the structure from the elements and provides a means to blend into the landscape. The thermal mass is less in the earth-bermed home, and the lack of earth cover does not allow for the thermal lag associated with the earth-covered roof. This means more severe temperature fluctuations in the earth-bermed house. The cooling potential is also reduced because less earth is in contact with the building and there is no vegetative cover to shade and provide transpirational cooling. Ken Labs, an architect and researcher specializing in the design of buildings that are responsive to climatic factors, estimates that 1200 square feet of earth-covered roof can provide a cooling potential of 1.5 million Btu. The conventional roof will require more maintenance and provide less protection from storms and external noise. Because more of its structure is protected from the elements, the earth-covered roof should last longer than the earth berm's roof. However, the conventional roof of the earth berm will cost less initially (some estimates claim that an earth-bermed house will cost 20 percent less than an earth-covered one). More will be said about this important issue in later chapters.

Amount of Earth Cover on Roof

The amount of soil on the roof (a design decision, see chap. 4) may range from an 8-to-10-inch-thick sod roof to a hill that is 9 feet deep or more. The structural systems used to support these earth loads (which are approximately 100 to 200 pounds per square foot of roof for each foot of earth cover) vary from poured concrete to highway culverts, with many alternatives in between. Roof soil depth depends on economic considerations and the climatic conditions of the region. Generally, it is best to put at least 3 feet of cover over the roof to get the moderating effects of the soil. Weekly variations in temperature will probably penetrate around 20 inches (although the exact thermal characteristics of roof soils are not yet well understood). The growth requirements of roof vegetation also affect the amount of earth cover. Most grasses and ground covers will require at

least 12 to 18 inches and small shrubs, 30 inches. Larger shurbs and trees will require depths of up to 60 inches of soil cover. Again, more will be said about this topic later.

Advantages of Earth Sheltering

Table 2-2 lists the energy advantages associated with the soil and vegetative cover of an earth sheltered building over a conventional building. Indoor temperatures of aboveground structures depend on the outside temperatures, which may vary up to 90° F. Subterranean temperature fluctuations, however, are much less.

TABLE 2-2
Energy Advantages

1. **Soil Cover Advantages**
   — smaller temperature differential between inside and outside of house
   — heat extraction in summer by cooler soil next to house
   — thermal lag effect
   — thermal mass effects

2. **Vegetative Cover Advantages**
   — shading effects and reduced heat gain during summer
   — increased insulation
   — cooling by transpiration and evaporation

3. **Other Benefits**
   — better soundproofing
   — reduced maintenance costs through energy conservation and structural durability
   — fire protection
   — protection from storms and earthquakes
   — increased security
   — environmental conservation through double use of land; more open space; increased water retention and less runoff; better oxygen and carbon dioxide exchange; more space for food production

---

Ground temperatures respond only to seasonal changes, and these responses occur only after considerable delay. At depths of 17 to 26 feet below the surface, the temperature is virtually constant and approaches the yearly average for the surface temperatures. At depths of 6 to 20 inches, the soil does not reflect the daily changes above ground. Research in Kentucky indicates that a three-month lag or phase shift in soil temperature occurs at a depth of 10 feet. This equals a lag behind the surface of about one week per foot of soil depth. Although earth sheltered homes usually are not buried deeply enough to take full advantage of this phase shift, some of the winter coolness and summer heat lags will benefit the temperature performance of the house.

The earth, then, is like a thermal blanket that warms and cools as well as protects, smoothing out the daily temperature variations on the surface and giving the buried house a more gentle environment in which to operate. Although earth is not a good insulator, it is an effective isolator and moderator. Figure 2-4 illustrates this phenomenon, using a creature with long experience in earth sheltering.

The temperature performance of earth shelters will, of course, vary with soil type and moisture, latitude and elevation, surface color, angle of slope, vegetative cover, and general climatic conditions. When more is known of these phenomena, it may be possible to tailor a home to take full advantage of these natural temperature moderators and eventually even to modify the moderators.

Table 2-3 gives the annual maximum and

2-4. Generalized daily and annual temperature fluctuations above and below ground.

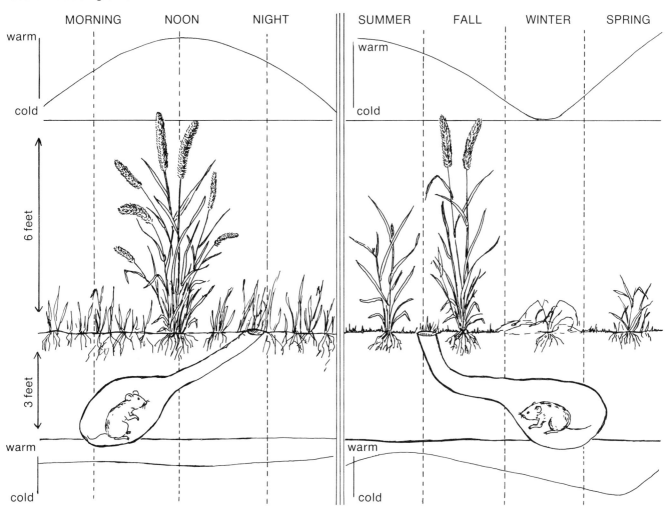

minimum earth temperatures at various locations. In summer, temperatures in a typical earth sheltered home in the Midwest theoretically should range from around 73° F near the top of its walls and roof to near 65° F around the floors. Winter soil temperature at the top of the walls should be approximately 45°F and about 55°F at the floor. These conditions are much more gentle than those on the surface and offer substantial heating and cooling benefits. In Lexington, Kentucky, the number of heating degree days was reduced by 43 percent, and the need for mechanical cooling was reduced to zero by going underground. As architect Malcolm Wells appropriately stated, "Putting soil on a house is like pulling a protective blanket over its shoulder."

TABLE 2-3
Average Maximum and Minimum Air and Earth
Temperatures (Integrated Average over Soil Depth of
2-12 feet; in Degrees Farenheit)

	Maximum (July)		Minimum (January)	
	air	earth	air	earth
Atlanta	87°	68°	33°	59°
Boston	81°	56°	23°	47°
Chicago	84°	56°	17°	46°
Dallas	96°	74°	34°	65°
Denver	87°	56°	16°	50°
Houston	94°	79°	42°	72°
Kansas City	88°	60°	18°	50°
Los Angeles	75°	68°	45°	66°
Miami	89°	78°	59°	72°
Midland (TX)	95°	70°	29°	65°
Minneapolis	82°	53°	3°	42°
New Orleans	90°	73°	44°	66°
Phoenix	105°	72°	38°	67°
Salt Lake City	93°	56°	19°	50°
Washington	88°	60°	28°	51°

REDUCED HEAT LOSS

Another important characteristic associated with earth sheltered buildings is the effect of earth cover and isolation on heat loss caused by infiltration. The earth sheltered house has fewer cracks exposed to the wind, and the earth cover deflects the wind up and over the house if the building is properly sited. This tightness and protection allow for control of ventilation and fit well with such adaptations as heat exchangers (ventilators that reclaim heat while exchanging inside and outside air) and earth tubes (pipes buried in the ground deeply enough so that the air passing through them is modified by ground temperatures).

INCREASED THERMAL MASS

The thermal mass of a house allows for absorption and retention of heat from internal sources and from solar radiation. Aboveground houses (even if constructed of heavy materials) store relatively little heat when compared with earth sheltered homes. A common complaint in passive and active solar homes relates to their limited ability to store heat over extended cloudy periods. The thermal mass of both soil and house materials act together in the earth sheltered home to maintain comfortable temperatures for up to a week with little or no external energy input.

COMFORT

The heat environment of the earth sheltered home is determined by mean radiant temperatures (MRT). The interior of a passively heated home is perceived to be more comfortable than the convectively conditioned environments of conventional structures. Radiant heat is sensed as a "warm" heat rather than the "cold" heat of forced air systems.

PROTECTION FROM POWER OUTAGES

The heat absorbtion and retention of earth sheltered homes is especially beneficial in times of power outages. Water pipes do not freeze, and the occupants remain comfortable because of the constant temperatures promoted by the thermal characteristics of the building and its earth cover (the temperature drop is less than 2° F per day). This constant temperature regime is also important to utility companies because a community of earth sheltered homes will not immediately increase power demand in periods of extremely hot or cold weather. Since electric rates are set by peak demand loads, earth sheltered homes should allow for more smooth and predictable demand curves and thus lower utility rates.

SUMMER COOLING

In contrast with conventional homes, earth shelters have cooler earth temperatures, lower transmission and infiltration losses, and higher thermal mass. These features are of substantial value for summer cooling. In appropriate regions of the country, this becomes one of the most obvious advantages of earth sheltered buildings. As electric rates soar, so do the bills for air conditioning. In earth sheltered homes, if air conditioners are needed at all, they are much reduced in size and cooling capacity.

IMAGINATIVE DESIGN

Creative use of the sun and earth are possible in the design process. Architects and engineers, limited only by their own energy and imagination, can provide warmth by the sun and cooling by the earth. Passive solar gain can heat the soil in the summer for winter use, or ice can be stored underground for summer cooling. Trombe walls (black, glass-covered concrete structures), thermal chimneys, and ventilating skylights, along with drainage pipes and earth tubes, can be used to condition incoming air.

ECOLOGICAL GAINS

As stated in chapter 1, the well-designed earth shelter will effectively integrate with the natural landscape. According to Thomas Bligh of the Massachusetts Institute of Technology, earth-covered housing developments can have two to three times more green space than conventionally developed areas. Earth-covered roofs capture and hold water, thereby aiding in the recharging of local aquifers. Added vegetation (especially if it is native) can increase the amount of wildlife habitat (food and shelter) and, regionally, can help reduce negative climatic effects (disturbed rainfall patterns, air movement, and temperature and humidity levels) of paved surfaces. In fact, the reflective

characteristics of the earth itself could be returned to a more natural condition with widespread underground construction.

Marginal land areas and wastelands can often be used and reclaimed by earth sheltered construction. The noise-reducing characteristics of earth sheltered structures have allowed them to be built in highly urbanized areas near freeways and airports. In addition, steep slopes, which cannot support conventional homes, will accommodate earth shelters.

However, it must be stated that the ecological benefits will not come automatically. Earth shelters can be as negative environmentally as aboveground construction if no thought is given to plant and animal distribution, water run-off, nutrient cycling, and myriad other ecological considerations. Environmental architecture, although simple in concept, may be challenging to implement.

PROTECTION FROM DISASTERS

Since many earth sheltered buildings are substantially reinforced and constructed with heavy materials, they protect from such disasters as fire, flood, wind, hail, tornadoes, and earthquakes. The structural shell is often a nonflammable, material of high mass such as concrete. With care in the choice of interior and exterior finishing materials, the fire hazard is substantially reduced over conventional housing (with the potential of lower insurance premiums). Where floods pose a threat, earth berms, gravity drains, and dry wells are useful in diverting and draining excess water. During a storm, the earth cover serves to deflect winds and hail and to distribute the impact of falling trees and other objects.

If the walls and roof are securely connected and supporting structures are properly spaced, the heavily reinforced structure can probably resist the forces of tornadoes and earthquakes. The large amount of glass on the face of an earth sheltered home represents the greatest hazard in a tornado. Small, windowless rooms on the north side are the best places for refuge. In areas prone to tornadoes or earthquakes (see fig. 2-5), this eventuality should be planned into the siting and design of the house.

Tornadoes and common storm tracks

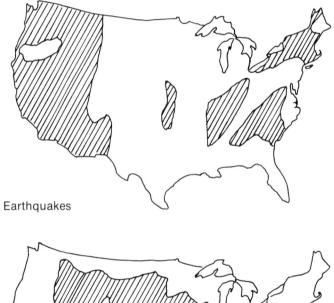

Earthquakes

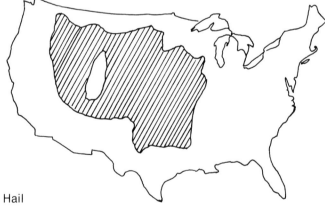

Hail

2-5. Areas of significant risk for tornadoes, storms, and earthquakes in the United States. (Source: Underground Space Center, 1982)

SOUNDPROOFING

The large earth mass around an earth sheltered home dampens the noise and vibrations associated with our daily lives. This is most useful in urban situations or sites close to highways, airports, or railroads.

REDUCED MAINTENANCE AND LIFE CYCLE COSTS

A very real benefit of earth sheltered homes is the low maintenance and life cycle costs. If durable products of good quality are used, the benign conditions underground will allow the materials to last almost indefinitely. To maximize this benefit, the exposed elements of the house should also be made of low-maintenance materials.

Possible Limitations

Earth sheltered homes are different from the architecture that we are accustomed to building, buying, and selling. For this reason it may be more difficult in some situations to find suitable building sites, competent professional help, or adequate financing.

REGIONAL DRAWBACKS

Earth sheltering performs better in some regions than in others, a condition to be expected for any climate-sensitive housing. Areas characterized by climatic extremes and low humidity will benefit most. In mild and humid regions, earth-covered structures may be inappropriate. Earth contact, a strategy that can be used in many situations in many different ways, must be incorporated into a design for a reason other than being perceived as a cure-all. More will be said about climatic factors in chapter 3.

SOCIAL AND LEGAL CONSTRAINTS

Some people have adverse reactions to anything "different" in the neighborhood. This reaction may be simple prejudice or ignorance of the benefits and advantages of energy-efficient housing. The human element can be a powerful force with the threat of court action involving codes and zoning.

Most aspects of earth sheltered housing are compatible with existing building codes. Some codes, however, may present obstacles to the further development of earth sheltered housing. These concern fire safety and egress, the effects of grade changes on safety, and the provision of natural light and ventilation. Since it is a frustrating and expensive proposition to be at variance with a code, it pays to design an earth shelter for compliance except in cases where compliance would compromise the structural or thermal integrity of the building. Following is a brief discussion of major problem areas. Chapter 4 will discuss the role of codes in the design process.

Every sleeping room is required to have at least one operable window or exterior door for emergency egress or rescue. For earth sheltered residences, it is suggested that windows that open to greenhouses or atriums fulfill this requirement. Overall, however, providing egress does not seem to be a major problem, although it demands attention during the planning stages.

Some building codes may require a guardrail to prevent children or others from falling off the earth shelter's roof. The codes are somewhat ambiguous on this point, and an exemption may be possible based on justifiable circumstances. It is probably best to include some kind of barrier into the design that is both appropriate and aesthetically pleasing.

Well-designed earth sheltered homes do provide adequate light and ventilation and should not be in conflict with the codes. Natural light in every room, however, may not be possible, and this code provision may have to be challenged. It is also recommended that in some instances artificial light be allowed as a substitute for natural light. Most codes require a glass area equal to 8 or 10 percent of the floor area in each habitable room. This can usually be met by incorporating skylights, arranging an open room around each window, or having rooms open into greenhouses or atria.

Any zoning ordinances prohibiting below-grade space from being considered habitable space will cause problems. Applying these rules to a well-designed earth sheltered residence violates the intent of zoning ordinances, which is to protect the general welfare of the community, and will have to be challenged. Often, these ordinances are a reaction to the "basement" houses of post–World War II days and are no longer relevant. Therefore, earth sheltered space may have to be defined separately. Earth sheltered homes are not unhealthy, substandard homes but can be attractive residences that will not bring down property values or detract from community aesthetic standards. This applies also to requirements for minimum height and floor area. Figure 2-6 shows an earth sheltered home in a typical community.

Where ordinances require a certain amount of "building-free" space (set backs or maximum lot-coverage requirements), it will have to be shown that earth shelters can be built with adequate open space for access by fire departments and utilities and for maintenance between buildings, without negatively affecting neighboring buildings, streets, and sidewalks during excavation. Earth-bermed houses may be more of a problem in this regard than earth-covered residences because the berms are not usable open space. Before buying a lot—and certainly before beginning construction—consult with local zoning officials and ascertain the attitudes of the neighborhood.

DESIGN CONSTRAINTS

Technically, earth shelters require more land than conventional homes because they are essentially horizontal in orientation. This may be a problem where building lots are small, and property setbacks are rigidly enforced.

HIGHER COSTS

The initial cost of an earth sheltered home may be higher than a conventional home (although keep in mind the life cycle benefits stated earlier). A 10–35-percent higher figure is often quoted, because of the high quality of required materials (structural, waterproofing, and insulative). In addition, excavation and added energy systems (greenhouses, passive solar windows and walls, and so on) raise costs above

2-6. Conventional-looking earth sheltered home. (Courtesy of Lon B. Simmons)

other energy-efficient homes such as superinsulated types.

The considerable confusion that exists concerning intital, operating, and life cycle costs involved in earth shelters is due in part to the relative lack of research on actual earth sheltered homes and to the uniqueness of the concept. If costs are nearly 50 percent higher than for conventional building, as claimed by some experts, the economics of the situation will exclude all potential homebuilders except the wealthy. However, if costs can be controlled by a combination of owner involvement, increased expertise by professionals, and cheaper building materials and methods, the benefits of earth sheltering will be available to many people.

It will take at least ten years before the added costs of building an earth shelter are returned to the owner by low operating and maintenance costs. The aboveground alternatives (superinsulated homes in particular) are thought to be more cost effective over the short term (ten years or less). However, with energy costs escalating, it is difficult to be definitive on this issue. Earth sheltering is probably the best investment for the long term, and, if the resale potential of earth sheltered residences is good, the short-term disadvantage may also be lessened.

DIFFICULT RESALE

Any unconventional home may have a problem with resale. Our society forces most of us into a rigid pattern of conformity in house design, although there are signs that this is changing in the face of higher energy bills. Houses that are aesthetically pleasing as well as energetically and ecologically appropriate will always find a buyer. As experience is gained in the earth shelter market, the question of resale potential should be more easily addressed.

Once a good site is found (with good solar exposure, soil conditions, and so forth—see chap. 3), it may be difficult to locate a knowledgeable architect, engineer, and contractor. Research by the potential homeowner is also necessary, and time must be taken to read, talk to people, go to seminars, and interview profes-

sionals. Without adequate preparation, an earth shelter homeowner may wind up paying the architect and contractor excessive amounts to cover costs and to allow for on-the-job learning.

Table 2-4 lists some inherent design considerations that may limit the use of earth sheltered techniques. Unknown phenomena concerning soil temperature changes and heat flow through walls and earth in differing soil types and moisture conditions make definitive answers difficult. Such variants as vegetative or snow cover, for example, can have significant effects. Designers must take into account the varying response times of thermal mass, seasonal lag in temperatures, eventual warming of the earth cover, placement of insulation, and depth of soil on the roof. Simple rules of thumb may be inadequate in the face of such complex-

TABLE 2-4
Potential Difficulties for Earth Sheltered Housing

— in predicting overall performance because of unknown behavior of heat flow through soil.

— in predicting the response time of the structure and earth cover to unusual situations such as temporary periods of overheating or overcooling.

— in designing an earth shelter to make use of ground temperatures. Those regions requiring heat have the coldest ground temperatures, and those requiring cooling have the highest. Also, the house itself warms the soil, and the effects of this are difficult to predict.

— in designing an earth shelter to take advantage of the thermal lag in seasonal soil temperatures. The walls of the home may be too cold in the spring, too warm in the fall.

— in designing an insulation scheme which can provide for heat retention in the winter and also allow for cooling in the summer.

— in providing for the control of humidity and condensation. Cool wall temperatures may provide cooling but promote condensation in areas with high outdoor humidity. Also, the earth covered building may be more difficult to ventilate with passive means (natural breezes).

— in providing enough soil cover on the roof for maximum use of soil temperature moderation and for the growth of plants. It becomes more expensive to provide the structural strength to support the earth loads necessary to achieve the maximum temperature benefits of soil cover and to grow plants without irrigation.

— in finding qualified personnel, experienced in earth sheltered construction.

— in making repairs

ity, underscoring the need for some professional assistance.

Other problems associated with earth sheltered or underground architecture concern dampness and mildew formation. If humid air makes contact with a wall with a surface temperature below the dew point, water will condense on the wall. If air circulation is poor, the water will remain, with subsequent growth of mold and mildew. For this reason, the insulation and ventilation systems are extremely important. In some cases the addition of mechanical systems is necessary. These can be small air conditioners or heat pumps (such as the water-heating type that can dehumidify and cool as well as heat water).

REPAIRS

If a leak occurs, the costs (and emotional stress) associated with its location and repair can be formidable. Usually, some soil cover and insulation must be removed and replaced with appropriate waterproofing (with the patient monitoring of performance during the next rain).

EXPERIENCES: ENERGY-EFFICIENT HOUSING

Reliable information on the actual performance of energy-efficient homes is just now becoming available. Well-designed superinsulated conventional homes, passive retrofits, new passive solar and earth sheltered homes, all efficiently reduce energy expenditures for winter heating. Some active solar homes perform well, but others have proven unsatisfactory. This also applies to double-envelope homes, which seem to be in an early stage of development, with only a few quantitative or computer studies giving an idea of how they compare to other energy-conscious designs.

Earth-Covered versus Earth-Bermed versus Aboveground: Performance Data

Computer models are used to evaluate more objectively the performance of the options shown in figures 2-1, 2-2, and 2-3. Researchers for the U.S. Department of Energy found that earth sheltered dwellings could be 33–58 percent more efficient in energy use than comparable aboveground conventional homes, depending on location. An Underground Space Center computer predicted the performance of a well-insulated aboveground home, an earth-bermed home, and a three-year-old earth-covered design (three years are required to stabilize the earth temperatures around the earth-covered house), all located in Minnesota, producing the results shown in table 2-5. The earth-covered design outperformed the others in both winter and summer, actually producing an excess of energy in the winter and requiring only a small amount of energy for summer cooling. The earth-berm model nearly matched the earth-covered design in winter performance but lagged far behind in summer energy requirements. However, it had only half the cooling requirement of a conventional home. The conventional home, although doing well, fell short of its earth sheltered counterparts.

TABLE 2-5
Computer Predictions of Home Energy Requirements

Type of House (Minneapolis Area)	Winter Heating (kwh)	Summer Cooling (kwh)
Well-insulated, conventional, without basement	6589	4219
Well-insulated, conventional with basement	3623	4672
Earth-berm (single story with conventional roof)	none (excess of 112)	2326
Earth-covered (20 inches of soil on roof)	none (excess of 313)	731

The above figures are computer predictions and thus subject to the limitations of the computer program. There are many such computer programs available but most are not able to calculate accurately the phenomena associated with passive heating, earth sheltering, and the details of climate variation and occupant lifestyle.

Regional homeowners' reports on actual home performance are presented in the literature (see bibliography), but these claims are difficult to compare with other homes because of variability in home design, climate, comfort expectations, and occupant lifestyle (now considered a major factor in energy use comparisons). These sources report savings of 50–80 percent in energy use for heating, with passive homes ranking higher in percent savings than active homes. Temperature and humidity ranges are reported as comfortable in all the energy-efficient homes (55°–85° F and 20–70 percent humidity). The lowest winter temperatures occur in aboveground homes and the highest humidities in earth sheltered homes (earth shelters made of wood report lower humidity levels than those made of concrete).

Of homes monitored and reported on by the Underground Space Center, Lawrence Berkeley Laboratory, Oklahoma State University, and the National Association of Home Builders, earth shelters proved to have the lowest energy requirements. According to these figures, conventional homes lost about 4–15 Btu per square foot per heating degree day, superinsulated homes lost .6 to 7.5 Btu, and earth sheltered homes lost in the range of .8 to 3 Btu. David Wright, leading passive solar architect and author, cites comparable figures from passive solar and earth bermed homes around the United States. A carefully instrumented earth sheltered home studied by the South Dakota School of Mines and Technology also bears out the validity of this data. Table 2-6 summarizes the performance data from these studies (not all home types were studied by every research group). It is important to note that earth shelters outperform aboveground homes when both cooling and heating needs—expressed as degree-day Fahrenheit (ddf)—are considered, as in the Oklahoma State University study.

Reports From Other Countries

Since earth sheltered techniques have been used for centuries in such countries as Greece, it is interesting to examine the results of recent

TABLE 2-6
Results of Energy Performance Studies for Homes in the United States

Housing Type	Source of Information	*Energy Requirement
Conventional (pre-1975)	Underground Space Center	4 to 15 Btu/sq. ft./hdd
Conventional (pre-1975)	Lawrence Berkely Labs	8 to 20 Btu/sq. ft./ddf
Conventional	National Association of Home Builders (NAHB)	10 Btu/sq. ft./ddf
Conventional	Oklahoma State Univ.	15 Btu/sq. ft./ddf
Moderately insulated Conventional	NAHB	7.5 Btu/sq. ft./ddf
New building practice superinsulated	Lawrence Berkely Labs, NAHB	5.0 Btu/sq. ft./ddf 0.6 to 1.1 Btu/ sq. ft./ddf
superinsulated	Lawrence Berkely Labs	1.6 Btu/sq. ft./ddf
passive superinsulated	Lawrence Berkely Labs	2.5 Btu/sq. ft./ddf
active superinsulated	Lawrence Berkely Labs	4.1 Btu/sq. ft./ddf
active solar	Lawrence Berkely Labs	3.8 Btu/sq. ft./ddf
passive solar	Lawrence Berkely Labs	1.8 Btu/sq. ft./ddf
passive solar	David Wright, A.I.A.	5.4 to 9.1 Btu/sq. ft./ddf
earth berm	David Wright, A.I.A.	5.3 to 7.3 Btu/sq. ft./ddf
earth-covered	South Dakota School of Mines and Technology	3.5 Btu/sq. ft./hdd
earth-covered	Minnesota—Underground Space Center	0.8 to 3.0 Btu/sq. ft./hdd
earth-covered	Lawrence Berkely Labs	2.4 Btu/sq. ft./ddf
earth-covered	Oklahoma State Univ.	6.2 Btu/sq. ft./ddf

* The energy requirements are expressed either in Btu per square foot per heating degree-day (hdd) or degree-day Fahrenheit (ddf), which includes both heating and cooling degree-days. The data from Oklahoma State University is total energy use and is the monthly range for five earth shelters for 1977 to 1978.

research from those areas. Yearly temperatures have remained stable in these ancient earth shelters (68°–84° F.) with humidity levels at 60–76 percent. Although the temperatures on the whole are quite acceptable, humidity is reported to be a problem, especially in the summer. In Australia, temperatures in dugouts range from 57° to 81° F. (mean radiant temperatures are around 65°–68° F.), with humidity levels ranging from 46 percent on dry days to 93 percent on wet days.

Incidentally, none of the countries that have a long history of people living in earth sheltered conditions report any negative health effects associated with the living conditions below grade. Evidently, the various sicknesses supposedly associated with damp atmospheres, poor air, and mildew are unfounded or exaggerated. More will be said about this later.

Negative Experiences

It is beyond the scope of this book to give a complete picture of all the problems associated with energy-efficient housing. This section will focus mainly on contrasts between aboveground or surface housing and earth sheltered construction with particular regard to reported negative experiences of homeowners.

No design, whether it be aboveground or below, will be without some problems, and those stated here should not necessarily be considered as inherent to a particular housing strategy. Often, the problem is a result of poor design or faulty construction and should not be considered a condemnation for the particular strategy used. In other cases, the design strategy itself is inappropriate for the climatic region or site. The challenge to future homeowners and designers is to differentiate between these two sources of error.

By examining the experiences of others, it may be possible to make better design choices for future homes. To this extent, a great debt is owed to those owners who, in the desire to help others, expose the problems they have encountered.

CONVENTIONAL HOMES

The energetic and ecological ills of the conventional home have been discussed in chapter

1. Briefly, they are: high heating and cooling bills because of poor thermal characteristics and complete dependence on fossil fuels; high maintenance and short life cycles; vulnerability to damage from natural and manmade disasters; dependence on mechanical systems for comfort control. To carry these drawbacks into the coming age of energy scarcity is maladaptive.

SUPERINSULATED HOMES AND DOUBLE-ENVELOPE HOMES

William Shurcliff, in his book *Superinsulated Houses and Double-Envelope Houses*, reports on ten superinsulated and ten double-envelope homes. He endorses the former and is very cautious about the latter. In general, the few negative reports about the superinsulated home focus on high humidity in winter, poor air quality, moisture formation in walls, little ability to use thermal mass and passive solar inputs, a sense of confinement, loss of space due to the added wall insulation, and continual reliance (be it ever so small) on conventional sources for air conditioning and heating. Since the home is aboveground, the durability of the exposed elements is also questioned.

The superinsulated home is a proven success, however, as long as enough heat can be provided by internal sources (stoves, refrigerators, people, and so forth) and cooling can be provided by nighttime ventilation or air conditioning. However, the design is based on the sealed environment—one that is separated from the outside surroundings and does not allow for environmental factors to make a contribution to physical and psychological comfort. Ecologically, it has most of the same drawbacks as any conventional home.

The double-envelope home is still in an early stage of development. Shurcliff lists the following concerns: loss of heat through the extensive glazing; cold basement or crawlspace; the need for added thermal mass; much inaccessible space (the envelope and crawlspace); fire hazards in the airspace (envelope); the required ventilation of large amounts of hot air, the need for extensive cooling on hot days; and the possible higher building costs than other alternatives such as the superinsulated home.

David Wright and Dennis Andrejko, in their book *Passive Solar Architecture*, found many of the same problems in a double-envelope home in North Carolina, as well as excess moisture and mildew buildup in the envelope. The problems may be solved through design alterations, for there are owners of double-envelope homes who report extraordinary success in attaining comfortable living conditions. Ecologically, these homes challenge the architect to blend them into the landscape because of high visability.

ACTIVE AND PASSIVE SOLAR HOMES

The major recorded complaints about active solar systems concern the added costs, the potential for breakdowns caused by the complexity of the systems, and disappointing performance. Although some systems work well, others have problems with delivering heat from storage areas to living spaces in the winter and with overheating in the summer. Some homes actually report increased fuel bills after installation of active panels. Although the potential for active solar is substantial, in recent years it has taken a secondary role to passive solar systems because of concerns related to operating reliability, dependence on solar firms for service, and possible damage to collectors by storms, dirt, and vandalism.

Solar Age magazine conducted a poll of seventy-one owners of passive solar aboveground homes and published the most frequent complaints. They dealt with fading of carpets, furniture, and drapes; overheating; underheating on cloudy winter days; large daily temperature swings; and lack of privacy. Because passive techniques are integral parts of many homes, it is difficult to attribute these complaints categorically to the passive solar system.

Active solar retrofits of conventional homes have presented some of the same problems as new active solar residences. If the house has not been well insulated, the active solar systems will not perform up to potential. The same

is true for passive solar retrofits of conventional homes, although the simpler nature of passive systems makes them more reliable and durable than active systems.

Solar homes are similar in style and appearance to conventional homes. Thus there is little reason to expect difficulty with financing, building codes, or zoning requirements. The solar homes, because they require exposure to the sun, have had some problems securing their "solar rights" indicating the need for careful site selection for these residences. In general, however, public response to these homes has been favorable.

EARTH SHELTERED HOMES

Because earth sheltered construction is an emerging technology, practical experience with actual homes is still in short supply. Earth sheltered buildings are by nature different from surface homes and thus cannot be built with the same objectives in mind. Many aesthetic problems have arisen because designers and owners have desired an earth shelter to appear to be something it is not—a conventional home. For this reason, some earth sheltered residences are mismatched with the surrounding natural and manmade environments.

Potential technical difficulties, to be covered at length in later chapters, include: heat loss through thermal leaks or wicks in insulation, cracked walls (especially parapets and retaining walls) from soil expansion, and water leaks from poor structural and waterproofing design. Also plant growth problems caused by inadequate soil cover and drainage have been reported, as have overhang and shading malfunctions. Very important shortcomings have been documented, involving transitions between buried and surface elements, especially in earth-bermed homes. Some homes have had structural problems such as roof deflections or problems with vent pipes, ducts, flues, windows, and skylights. Humidity and condensation can also damage the home.

Problems with building codes, zoning ordinances, financing, and public acceptance are often resolved through the education of the officials and persons involved. Regrettably, however, there have been cases where no solution was found and construction was halted.

While most of the technical problems mentioned above have readily available solutions, one serious problem remains—high construction costs for earth sheltered homes. Two studies for the U.S. Department of Energy have concluded that earth sheltered homes will cost 23–36 percent more than comparable surface houses, and, as previously mentioned, energy savings by earth sheltered homes are not expected to offset higher costs within a period of less than ten years. However, owners of earth shelters and some architects disagree with these figures, and most reports in the literature quote costs for earth sheltered structures that are approximately 10 percent higher than conventional. (See Appendix D for more information on costs).

To summarize, aboveground homes do not at present have the potential to provide adequate cooling by natural means. Because they are on the surface, they also have difficulties with durability, maintenance, and ecological soundness. Conventional homes also lack the protection necessary to survive natural and manmade disasters. Earth sheltered homes, although subject to some serious shortcomings, maximize the unique opportunities for energy conservation and ecological landscaping. The higher construction costs of earth sheltered structures is an area of controversy that requires close inspection before the decision to build is made.

Firsthand Experiences

Having lived in two conventional homes, a suburban ranch house and a rural farmhouse, we moved into our earth sheltered home in 1979 (see fig. 3-11). We found a near perfect site located about a half mile from our nearest neighbors, who were informally queried about their opinions of earth shelters. Located on fifteen acres of gently sloping prairie in south cen-

tral Kansas, it is in an area classified on climatic charts as having hot, moderately humid summers and cold-to-moderate winters. In other words, it is a land of extremes and an appropriate area for earth sheltered architecture.

After extensive research, which included seminars and interviews with earth shelter architects, designers, and homeowners, we formulated the design features we wanted, and we looked for and found appropriate professional personnel to help us accomplish our goals. Local building codes and zoning ordinances, which actually did not apply because we were building in a rural area, were still followed. Financing was arranged through private means, and the neighbors were found to be enthusiastic about having an earth shelter built nearby.

Our expreiences building the house are documented throughout the book. After moving into it, we systematically recorded its thermal performance and economic and ecological impact (see Appendix A). When our energy expenditures in the earth sheltered home were compared to those in the rural farmhouse, we made the following conclusions.

Electrical use: the earth shelter used 30 percent less even with more people using the home for longer periods. Most of the savings came from the absence of air conditioning and mechanical heating.

Fuel use: the earth shelter used 67 percent less wood (a half cord per winter) and 100 percent less propane (no furnace) in the heating season. The passive solar gain and earth sheltered benefits (less heat loss) were responsible for this difference.

Comfort levels: in the year immediately following construction, the performance of the home was not as good as in later years. After the soil temperatures stabilized and shade was provided, temperatures ranged from 67° to 72° F in the winter with an auxiliary input of a half cord of mixed hardwoods per heating season. Winter humidity levels were between 30 and 50 percent. Summer temperatures were between 75° and 80° F, and humidity ranged from 40 to 68 percent. Winter conditions were very comfortable whereas summer conditions could be described as acceptable. The comfort level of air conditioning was not achieved as effective temperatures in summer approached the edge of the comfort range (see chap. 3). However, with moving air and light clothing, we were just as comfortable as neighbors who set their thermostats at 80° F to avoid astronomical air-conditioning bills.

Temperature ranges (no auxiliary input): the lowest temperature during an unoccupied thirty-day period in January 1982 was 62° F. The highest temperature reached during an unoccupied sixty-day period in June and July 1983 was 78° F. In each season, the outside temperatures were characteristic of cold Kansas winters (below 0° F) and scorching Kansas summers (above 105° F).

From our experiences as they relate to the limitations mentioned in table 2-4, the following conclusions can be reached:

Predicted performance: the architect's sizing of the solar collection systems (windows and Trombe walls) appeared to be correct, for winter temperatures ranged between 67° and 72° F. Temperatures were maintained (a 3° drop) by the thermal mass for approximately three to four days in cloudy weather. With external shutters, the summer temperatures were acceptable. Without shading, the house became too warm (87° F). I fully expect the increased shading provided by deciduous trees and vines planted on the south side of the house to bring temperatures well within the comfort range.

Slow response of thermal mass: with solar input, this was not a problem. After an extended cloudy period, the thermal mass required about a tenth of a cord of wood to produce enough heat to raise the mean radiant temperature back to sunny conditions.

Seasonal lag in soil-wall temperatures: in the spring are occasionally cooler than the outside daily temperatures because of the lag of the winter temperatures into the month of April. With solar input, the difference is not important. If little solar input is available, then auxiliary heating is required to keep interior temperatures in the 70° range. If temperatures around 68° are acceptable, this phenomenon is

of little consequence. In the autumn, the lag of summer temperatures occasionally causes the wall surface temperatures to be higher than the outside temperatures. In this case, simply opening the windows restores an equilibrium. Temperature differences are small in each of the above situations, and most people will not be inconvenienced by this lag phenomena. In fact it is often an advantage where spring temperatures are too warm or fall temperatures too cold. At no time were we unable to take advantage of cooler night temperatures because of the daily thermal mass swing. If natural ventilation did not provide the cooling, an exhaust fan was used at night. The next day, the home was closed to hot outside conditions, and the temperatures never exceeded 80° F.

Inhibited summer cooling by insulation: the walls and roof of the house are insulated on the outside, causing them to be somewhat isolated from the cooling effects of the adjacent soil. The interior surfaces of the walls and roof have never exceeded 81° F, however. The floor slab is not insulated and provides substantial cooling as the deep soil layer beneath has temperatures around 64° F.

Humidity and mildew: condensation did form on lightly insulated walls of a closed-in corner of the garage. When ventilation was provided, the problem was solved, demonstrating the importance of adequate ventilation. No other rooms in the home have had this problem.

Vegetation on roof: during a period of drought, the grass on the roof turned brown but did not die. Because they have the ability to adapt to local drought conditions, native grasses (buffalo and bluestems) were used. The soil depth on our roof ranges from 3 feet in front to 2 feet in back, which is probably a minimum for supporting vegetation without irrigation in our area (average annual rainfall of 32 inches).

Contractors: we found local help to be adequate to the task as long as the architect and engineer provided close supervision and guidance through regularly scheduled visits. Well-detailed blueprints were essential.

Repair work: some leaks did occur around the skylight and vent pipes. These were repaired by the contractor and paid for by the architect.

Costs: the final cost (including special materials and procedures for an expansive clay soil) was comparable to the cost of building a custom aboveground home in the same area. However, I was extensively involved in the construction, and the fees charged are considerably higher now. Tax credits for applicable solar and conservation measures have returned some of the costs, as have property tax refunds for energy-efficient homes. Low operating and maintenance costs are saving money on a continual basis. Our savings are about $300 per year on heating and $200 per year on cooling. In four years we have spent no money for external repairs, in spite of three major hailstorms in that period. My insurance rates are about 20 percent lower than conventional rates because the house has a fireproof rating.

RECOMMENDATIONS

For people considering an earth sheltered home, a fundamental evaluation of motives is needed at the start. If energy conservation is the primary goal, other options such as the super-insulated or passive solar homes should be examined as well. If cooling, protection, and security are also important, then an earth sheltered home becomes the most desirable option. Table 2-7 provides a method for making a decision on whether to build an earth sheltered home. If the positive values outweigh the negatives, then an earth shelter becomes feasible. Otherwise, another approach may be the best choice. The decision, in reality, is much more arbitrary than this chart suggests, but using it should give a fair idea of the initial feasibility of the concept.

In most cases there will be no real estate agent in the beginning stages to assist in the acquisition and assessment of financing, site, plans, and builders. Therefore, any potential owner should become familiar with the concept of earth sheltering through study and communication with people in the field (see Appendix B).

TABLE 2-7

To Build or Not to Build

For each of the following factors, assign a positive or negative value according to how important that factor is to you. If the end total is negative, the building of an earth shelter will be more difficult and its feasibility should be questioned. If the factor is important, write down the highest value allowable, either positive or negative. Enter lower values if the factor is of lesser significance to your situation. An example is given for the author's situation in central Kansas. The total of +5 indicates that an earth shelter is feasible in central Kansas because the positive factors (those favoring earth shelters) outweigh the negative (those factors that make earth shelters difficult to build or inhabit).

Factor	Maximum Value Allowable Estimate	Example
Protection value from storms, disasters	+2	2
Protection value from noise	+2	1
Cooling potential	+3	3
Solar storage potential	+2	2
Long-term operation costs	+4	4
Lower maintenance	+4	2
Durability and life span	+3	2
Difficulty in securing financing	−7	−2
High initial cost of earth shelter	−2	−1
Coding or zoning problems	−3	0
High water table	−8	0
Soil problems (see chap. 3)	−4	−3
Ventilation problems (see chap. 3)	−3	−1
Humidity problems (see chap. 3)	−4	−2
Labor problems (no qualified personnel)	−7	−2
Construction problems (lack of materials)	−6	0
		TOTAL +5

Once the decision to build an earth shelter has been reached, it is necessary to make a personal inventory. First must come a personal financial assessment. No one should build a home that they cannot pay for. Next, the proposed building site should be evaluated, including an analysis of the neighborhood, zoning ordinances and services, as well as such physical characteristics as soil type and drainage.

Designs that are close to conventional in appearance will probably be received more favorably by loan officers. However, the special attributes of an earth shelter should not be compromised in this situation. Codes must be adhered to as much as possible and professional assistance should be secured, especially in waterproofing and in structural design. If a contractor is to be employed, it is good to find one with experience in the building of earth shelters, or the contractor should at least be experienced with the materials to be incorporated into the building.

More and more individuals prefer to build their own earth sheltered residences. For them, books, manuals, and training schools are available. Constructing any house is a major challenge but the earth sheltered house is especially demanding because of the need for a sound structure, specialized insulation, and effective waterproofing system.

Owner-builders can save a substantial amount of money (up to 80 percent by some estimates), but the headaches can be legion. For instance, local designers, code officials, lenders, subcontractors, and materials suppliers may be reluctant to work with nonprofessionals unless that individual has demonstrated skills in construction. Furthermore, obtaining approval of structural and waterproofing systems may be difficult unless building inspectors are assured that some professional was involved in their design and implementation. In the past, the choice of home has been an economic and emotional one. The average American is influenced by status, ego, conformity, "curb appeal," and cost. As Wendell Berry stated in his highly acclaimed book, *The Unsettling of America*, "The modern house is not a response to its place but rather to the affluence and social status of its owner. It is the first means by which the modern mentality imposes itself upon the world." With the growing energy and environmental awareness, this attitude is becoming archaic.

CHAPTER 3

SITE EVALUATION AND SELECTION

I will lift up mine eyes unto the hills from whence cometh my help.

Psalms 121:1

The eventual success of an earth sheltered home depends on how well the building is integrated with its environment. With careful planning, the natural energy of the site's microclimate can be used, reducing dependence on fossil fuels. If not, the building will fight the climate and be less efficient. In the words of Frank Lloyd Wright, "I think it far better to go with the natural climate than to try to fix a special artificial climate of your own."

Proper site integration often calls for the intervention of architects, designers, and engineers, and rather than a substitute for professional help, this discussion serves as a general guide.

Site evaluation and planning begin by determining whether the climate is conducive to the proper functioning of an earth sheltered home. Variations in air and ground temperature, precipitation, humidity, solar angle, cloudiness, and wind direction and force must all be considered.

PRINCIPLES: CLIMATIC FACTORS

The advantages of earth sheltering are impressive, but the degree to which these advantages are realized depends on climatic factors and design. Many climatic handicaps may be overcome through appropriate adjustments. For example, in areas that require ventilation to reduce the negative effects of humidity, wind scoops, vents, and mechanical systems (air conditioning) can be planned into the structure.

Climatic Regions

Figure 3-1 shows the climatic regions of the United States, and table 3-1 lists the climatic data most important to determining the potential problems for earth sheltering for selected cities. Those areas with large differences in high and low temperatures, cool summer soil temperatures, and low humidity and dew points will have the fewest problems.

The required amount of heating is expressed in heating degree-days (hdd) and of cooling in cooling degree-days (cdd). Both are calculated by subtracting the average daily outdoor temperatures from a base temperature (usually 65° F for hdd and 75° for cdd) and totaling the differences over the number of days in the heating or cooling period. For example, Kansas City has 5,357 base-65° hdd for a heating season. The daily average temperatures in the heating season for Kansas City were subtracted from the base temperature of 65° F, and these differences were added. Likewise, Kansas City

TABLE 3-1
Annual Climate Data Charts

HEATING City	Region	degree-days 68° F base	% hours less than 68° F	% hours less than 50° F	number of days with high temp. less than 32° F	record low temp. (°F)	% possible sun annual basis	% hours in comfort zone
Minneapolis	1	8,159	79.4	53	84	− 34	58	10.7
Chicago	2	6,127	73.7	48	45	− 23	57	12.5
Denver	3	6,016	79.1	49	21	− 30	70	9.3
Salt Lake City	4	5,983	75.5	50	24	− 30	70	11.0
Boston	5	5,621	79.6	47	27	− 18	60	10.8
Kansas City	6	5,357	64.2	40	28	− 22	67	13.6
Seattle	7	5,185	92.7	52	3	0	49	6.0
Washington	8	4,211	66.4	37	9	1	58	12.4
Atlanta	9	3,095	59.3	26	3	− 9	61	12.5
Midland	10	2,621	53.4	25	2	− 11	—	16.9
Dallas	11	2,382	48.8	21	3	− 8	68	12.7
Los Angeles	12	1,819	80.3	7	5	23	73	15.2
Phoenix	13	1,552	44.7	15	0	16	86	12.9
Houston	14	1,434	39.0	7	1	5	57	6.6
New Orleans	15	465	41.6	11	0	7	59	9.3
Miami	16	206	15.7	1	0	26	67	18.1

COOLING City	Region	degree-days 78° F base	% hours more than 78° F	% hours ventilation is effective	% hours requiring mechanical cooling	% hours requiring dehumidification	total % hours requiring nonpassive cooling	% hours passive cooling other than ventilation is effective	number of days with high temp. more than 90° F	record high temp. (°F)
Seattle	7	19	1.2	1.1	0.0	0.0	0.0	0.1	3	100
Los Angeles	12	28	2.0	1.5	0.0	1.4	1.4	0.4	0	110
Boston	5	127	6.0	5.3	0.2	3.5	3.7	0.3	12	104
Denver	3	145	7.1	3.1	0.0	0.0	0.0	4.0	33	105
Minneapolis	1	160	7.2	6.3	0.6	2.5	3.1	0.3	15	108
Chicago	2	241	10.0	8.5	0.4	3.9	4.3	0.7	21	105
Salt Lake City	4	276	10.2	4.5	0.0	0.0	0.0	5.7	58	107
Washington	8	322	14.2	10.9	1.5	8.0	9.5	0.5	37	103
Atlanta	9	397	17.2	14.2	0.5	12.2	12.7	1.0	22	103
Kansas City	6	496	17.9	14.1	1.2	4.6	5.8	1.9	40	113
Midland	10	731	23.7	15.2	0.1	2.4	2.5	8.6	92	109
New Orleans	15	772	33.5	19.2	1.6	28.0	29.5	0.3	68	102
Houston	14	998	40.4	22.9	0.3	31.2	31.5	0.0	88	108
Miami	16	1,045	50.2	35.4	0.6	30.0	30.6	0.1	30	100
Dallas	11	1,051	31.9	21.5	3.0	8.1	11.1	5.3	92	112
Phoenix	13	1,554	35.7	13.6	2.1	0.7	2.8	19.8	165	118

Continued

TABLE 3-1
Annual Climate Data Charts

JANUARY — City	Region	average temp. (°F)	average high (°F)	average low (°F)	average high minus low (°F)	ground temp. A 0–6 feet (°F)	ground temp. B 2–12 feet (°F)	ground temp. B minus average low	average daily solar radiation (Btu/sf) vertical, south-facing	average daily solar radiation (Btu/sf) horizontal surface	% possible sun	% relative humidity midnight	% relative humidity midday	mean wind speed	wind direction
Minneapolis	1	12.2	21.2	3.2	18.0	33	42	38.8	921	464	51	76	66	10.4	NW
Chicago	2	24.3	31.5	17.0	14.5	38	46	29.0	921	507	44	71	65	11.5	W
Denver	3	29.9	43.5	16.2	27.3	42	50	33.8	1,440	840	72	63	45	9.1	S
Salt Lake City	4	28.0	37.4	18.5	18.9	41	50	31.5	1,129	639	48	77	69	7.7	SSE
Boston	5	29.2	35.9	22.5	13.4	40	47	24.5	878	475	54	66	58	14.2	NW
Kansas City	6	27.1	35.7	18.4	17.3	42	50	31.6	1,098	648	58	71	64	10.7	—
Seattle	7	38.2	43.5	33.0	10.5	44	48	15.0	559	262	21	79	74	10.1	SSW
Washington	8	35.6	43.5	27.7	15.8	43	51	23.3	959	572	49	66	54	10.0	NW
Atlanta	9	42.4	51.4	33.4	18.0	52	59	25.6	1,041	718	48	75	60	10.5	NW
Midland	10	43.6	57.8	29.4	28.4	58	65	35.6	1,496	1,081	—	61	45	10.2	S
Dallas	11	44.8	55.7	33.9	21.8	58	65	31.1	1,164	821	56	74	60	11.2	S
Los Angeles	12	54.5	63.5	45.4	18.1	61	66	20.6	1,353	926	70	70	55	6.6	W
Phoenix	13	51.2	64.8	37.6	27.2	61	67	29.4	1,472	1,021	78	56	44	5.2	E
Houston	14	52.1	62.8	41.5	21.3	66	72	30.5	1,014	772	42	85	66	8.3	NNW
New Orleans	15	52.9	62.3	43.5	18.8	60	66	22.5	1,097	835	44	83	67	9.5	—
Miami	16	67.2	75.6	58.7	16.9	68	72	13.3	1,236	1,057	48	81	60	9.4	NNW

JULY — City	Region	average temp. (°F)	average high (°F)	average low (°F)	average high minus low (°F)	ground temp. A 0–6 feet (°F)	ground temp. B 2–12 feet (°F)	ground temp. B minus average high	average daily solar radiation (Btu/sf) vertical, south-facing	average daily solar radiation (Btu/sf) horizontal surface	% possible sun	July 21st minimum dew point temp. (°F)	July 21st average dew point temp. (°F)	July 21st maximum dew point temp. (°F)	% relative humidity midnight	% relative humidity midday	mean wind speed	wind direction
Seattle	7	64.5	75.1	53.8	21.3	59	54	21.1	405	2,248	65	49.6	52.9	55.7	67	49	8.4	S
Los Angeles	12	68.5	74.8	62.1	12.7	74	68	6.8	993	2,307	80	57.8	59.8	61.2	83	69	7.6	W
Boston	5	73.3	81.4	65.1	16.3	63	56	25.1	940	1,749	66	57.0	61.7	65.4	76	56	10.9	S
Denver	3	73.0	87.4	58.6	28.8	65	56	31.4	1,130	2,273	71	36.9	44.5	50.5	57	36	8.5	S
Minneapolis	1	71.9	82.4	61.4	21.0	62	53	26.4	1,140	1,970	71	58.8	62.9	67.6	75	54	9.3	S
Chicago	2	74.7	84.4	65.0	17.4	65	56	28.4	1,026	1,944	70	60.2	64.9	68.8	72	55	8.4	S
Salt Lake City	4	76.7	92.8	60.5	32.3	65	56	36.8	1,328	2,590	84	37.5	44.8	50.5	41	26	9.4	S
Washington	8	78.7	88.2	69.1	19.1	68	60	28.2	884	1,817	63	63.3	66.9	70.9	76	52	8.2	S
Atlanta	9	78.0	86.5	69.4	17.1	75	68	18.5	775	1,812	62	66.1	69.1	72.1	87	63	7.4	S
Kansas City	6	77.5	88.0	68.9	19.1	69	60	28.0	1,034	2,102	81	64.5	67.6	70.8	73	53	8.5	—
Midland	10	82.3	95.0	69.5	25.5	77	70	25.0	989	2,389	—	56.8	61.0	65.4	55	41	10.8	S
New Orleans	15	81.9	90.4	73.3	17.1	78	73	17.4	728	1,813	57	70.9	74.1	76.7	89	66	6.2	—
Houston	14	83.3	93.8	72.8	21.0	84	79	14.8	734	1,828	67	74.0	75.8	77.3	89	58	6.5	S
Miami	16	82.3	89.1	75.5	13.6	83	78	11.1	677	1,763	74	70.3	73.3	75.5	82	65	7.8	S
Dallas	11	84.8	95.5	74.0	21.5	81	74	21.5	893	2,122	78	64.4	68.5	72.1	67	49	9.4	S
Phoenix	13	91.2	104.8	77.5	27.3	78	72	32.8	1,059	2,406	85	56.1	60.7	65.6	33	28	7.2	S

(Source: Underground Space Center, 1982)

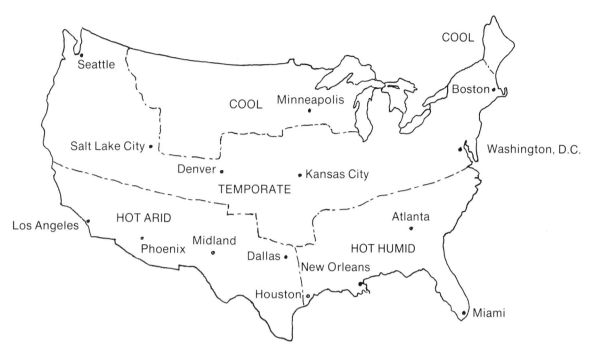

3-1. Climate regions of the United States showing location of cities in those regions.

has 496 base-78° days that are calculated similarly, except that a base of 78° is used for temperatures in the cooling season. The data for the months of January and July are especially useful since conditions in these months are the most severe.

Regional Problems for Earth Sheltering

According to regional maps compiled for use in determining the potential of earth sheltering for various areas of the country (see fig. 3-2), earth sheltering may be the best strategy for regions with high cooling and heating needs, high temperature differentials, and low humidity. For example, the region around Phoenix, Arizona, would be ideal, as would Salt Lake City, Utah; Medford, Oregon; and Seattle, Washington. Miami, Florida, and Houston, Texas, should experience less benefit from earth sheltering.

The southeastern regions of the United States are considered marginal areas for earth sheltering because of very high absolute humidity with summer dew-point temperatures near 70° F requiring extensive ventilation. There are also small day-night temperature swings, and ground temperatures are too high to permit earth-contact cooling. Earth-covered structures (soil on roof) may compromise the necessary ventilation strategies in this region unless proper design features are incorporated.

All climates present situations that can be used to good advantage by earth sheltering, but because of the difficulty in quantifying all the subtle daily and seasonal variations (let alone variations in altitude and local conditions), it is not possible to rule out definitively certain regions. The maps in figure 3-2 only identify possible climatic difficulties that must be considered in the design process.

Figure 3-3 shows an earth sheltered home in an ideal climate that allows the maximum use of earth contact. The extremes of temperature from day to day and season to season highlight the value of thermal mass and thermal lag (see chap. 2) associated with earth sheltering. Cool night temperatures and low humidity

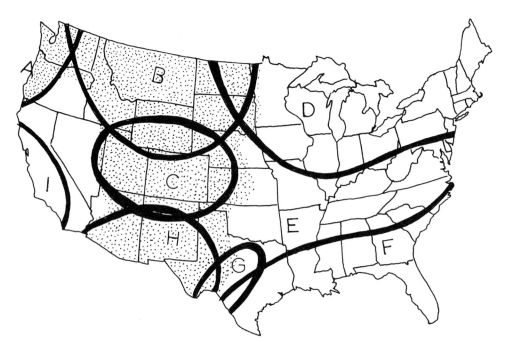

3-2. Regional suitability for earth sheltered homes (the best areas for earth sheltered homes are stippled). Zones A, B, and C: Cold, cloudy winters, and dry summers with cool soil temperatures favor earth-covered homes. Zone D: Cold winters and cool summer soil temperatures favor earth-covered homes but high summer humidity makes condensation a potential problem. Zone E: Hot summers and occasionally severe winters favor earth-covered homes but high summer humidity and warm soil temperatures place a premium on quality of design. Zone F: High summer humidity and warm ground temperatures may be problematic for earth-covered homes. The need for quick cooling by ventilation may favor other types of homes. Zone G: A transition area—the western parts favor earth-covered homes while the eastern sections pose some problems because of summer humidity. Zone H: Hot, dry summers and sunny winters favor earth-covering along with other passive design strategies. Zone I: Although earth-covered homes are appropriate for this zone, the mild climate negates the need for any extraordinary climate control strategy. (Source: modified from information provided by Ken Labs)

allow for the ventilation of the structure at night without substantially increasing the risk of condensation. Sufficient winds will allow for natural ventilation. Adequate winter sunshine permits the collection and storage of solar radiation for space heating. When summer soil temperatures are low, heat is drawn from the house, which has earth contact. In summary, reduced infiltration, the microclimate, thermal environment of the soil, and heat storage capacity of the thermal mass and earth are used to maximum advantage. Table 3-2 evaluates the advantages and disadvantages of earth sheltering on a regional basis and should help in deciding whether the benefits of earth sheltered homes match the needed climatic design priorities.

PRINCIPLES: COMFORT REQUIREMENTS

An understanding of human comfort depends on a basic knowledge of human physiology and principles of heat storage (temperature), heat transfer, relative humidity, and air movement. The human body possesses physiological mechanisms that strive to maintain a specified heat balance with the environment. Heat is produced by metabolism or muscular contractions (shivering), lost through heat transfer to the environment (for example, dilating peripheral blood vessels and sweating), and retained through heat-conserving mechanisms (such as constricting peripheral blood vessels and wear-

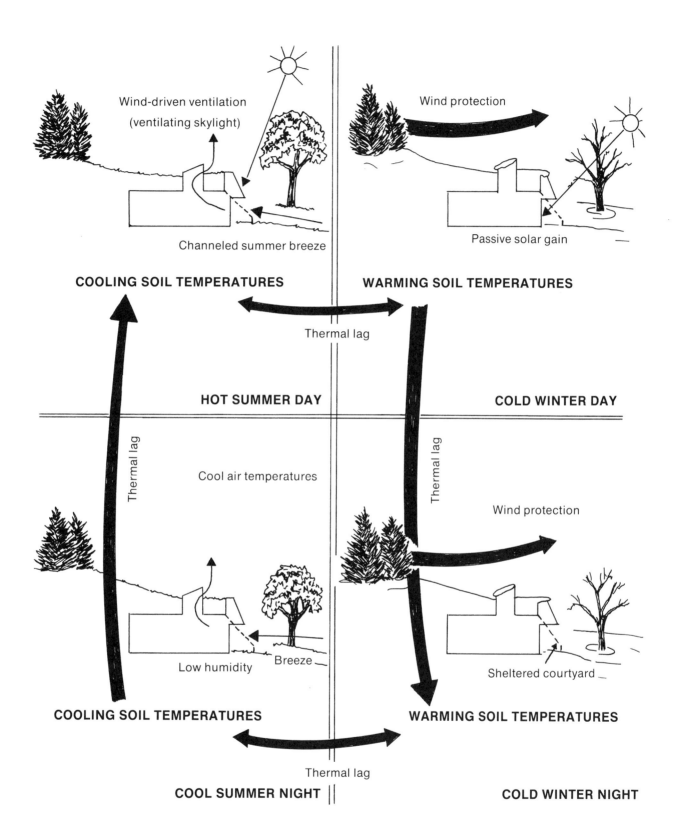

Wind-driven ventilation
(ventilating skylight)

Channeled summer breeze

Wind protection

Passive solar gain

COOLING SOIL TEMPERATURES

WARMING SOIL TEMPERATURES

Thermal lag

HOT SUMMER DAY

COLD WINTER DAY

Thermal lag

Cool air temperatures

Thermal lag

Wind protection

Low humidity

Breeze

Sheltered courtyard

COOLING SOIL TEMPERATURES

WARMING SOIL TEMPERATURES

Thermal lag

COOL SUMMER NIGHT

COLD WINTER NIGHT

3-3. Optimal climatic interactions for an earth sheltered home. Arrows indicate daily and seasonal transitions of soil temperatures.

TABLE 3-2
Regional Suitability

Region and Typical City	General Criteria: low maintenance	efficient land use	aesthetics	fire protection	earthquake resistance	tornado protection	Climate-Related Criteria: reduce winter heating	promote passive cooling	reduce mechanical cooling	Special Concerns: summer condensation	summer ventilation
1 Minneapolis	Moderately	Moderately	Important	Important	Slightly	Important	Very	Slightly	Slightly	Moderately	Slightly
2 Chicago	Moderately	Important	Important	Moderately	Slightly	Important	Very	Moderately	Moderately	Moderately	Moderately
3 Denver	Moderately	Important	Important	Important	Important	Not	Very	Slightly	Slightly	Not	Not
4 Salt Lake City	Moderately	Important	Important	Important	Important	Not	Very	Moderately	Moderately	Not	Slightly
5 Boston	Moderately	Important	Important	Moderately	Moderately	Not	Very	Slightly	Slightly	Moderately	Slightly
6 Kansas City	Moderately	Moderately	Important	Moderately	Important	Important	Very	Moderately	Important	Moderately	Important
7 Seattle	Moderately	Important	Important	Moderately	Important	Not	Very	Not	Not	Not	Not
8 Washington, DC	Not	Very	Important	Moderately	Important	Not	Important	Moderately	Important	Important	Important
9 Atlanta	Moderately	Important	Important	Moderately	Moderately	Important	Moderately	Moderately	Very	Very	Important
10 Midland	Moderately	Moderately	Important	Important	Slightly	Important	Slightly	Important	Important	Moderately	Important
11 Dallas	Moderately	Moderately	Important	Moderately	Not	Very	Slightly	Important	Very	Important	Very
12 Los Angeles	Moderately	Very	Important	Very	Very	Not	Not	Not	Not	Slightly	Not
13 Phoenix	Moderately	Important	Important	Moderately	Very	Not	Not	Very	Very	Slightly	Important
14 Houston	Moderately	Important	Important	Moderately	Not	Important	Not	Slightly	Very	Very	Very
15 New Orleans	Moderately	Important	Important	Moderately	Not	Important	Not	Slightly	Very	Very	Very
16 Miami	Moderately	Important	Important	Moderately	Not	Not	Not	Not	Very	Very	Very

Key:

Very Important

Important

Moderately Important

Slightly Important

Not Important

ing additional clothing). When the body is in equilibrium with the environment (heat production equals heat loss), we are comfortable; when heat gain and heat loss are out of balance, we feel discomfort.

Temperature is a measure of how much energy is stored by a substance. Different materials require different amounts of energy to be raised to the same temperatures. Air has a low specific heat requirement (the amount of energy needed to raise a given weight of a substance by 1° F), and concrete has a comparatively high specific heat. Air heats up rapidly; concrete heats slowly and retains its heat much longer. In this way materials composed of concrete (or some other high-mass material) can moderate temperature extremes in an area.

Heat energy can be transferred in space by radiation, conduction, convection, and changes of state. Heat always moves from a warmer substance to a cooler one, and temperature differences are eventually equalized. Changes of state consume vast amounts of energy. One hundred eighty Btu are needed to heat one pound of water (0.12 gallons) from freezing to boiling. Another thousand Btu are required to evaporate the pound of water. This explains not only how our bodies can lose heat by sweating and evaporation but also why we feel uncomfortable in humid conditions (the sweat does not evaporate as rapidly).

Relative humidity is the ratio of the actual amount of moisture in the air to the maximum amount the air can hold at a given temperature. When a relative humidity of 60 percent is present with a temperature of 80° F, we feel hot and sweaty because we cannot lose heat by evaporation very quickly. If air movement is increased or humidity decreased, the loss of heat by convection and evaporation is increased, and the temperature becomes more tolerable. Conversely, air movement with high humidity at low temperatures accentuates the impact of cold by promoting heat loss, producing a cold, raw feeling. In cold and wet climates, the winds can carry away heat and create a "wind-chill" factor. In hot and arid situations, air movement can dry out the body by carrying away precious water. Our bodies are adaptable to small variations, but our ability to survive depends mainly on our aptitude for building shelters that manipulate the environment to our advantage.

The American Society of Heating, Refrigerating, and Air-Conditioning Engineers (ASHRAE) has defined the comfort standards for American homes (table 3-3). Of course these standards are somewhat arbitrary and are influenced by the amount of clothing and activity of the subjects and their personal preferences. The British have defined an ideal temperature to be 66$\frac{1}{10}$° F, the Germans 69½°, and the Americans 73°, at approximately 50 percent humidity, with little air movement. Women prefer temperatures about 1° warmer than men, and old people prefer warmer temperatures than young people. Persons acclimated to a cold or warm temperature are better able to bear it than those just recently introduced to it. Obviously, discomfort occurs at different levels for different individuals.

The conventional house, like most such homes, is heated by a forced-air (convective) heating system and cooled by central air conditioning. The earth shelter is heated by radiation from the walls, ceiling, and floor and cooled by earth contact, which conducts heat from the interior to the soil. These two systems differ in how they influence the body's heat balance.

TABLE 3-3

ASHRAE Comfort Standards

Dry bulb temperature	73°–77° F
Mean Radiant Temperatures	70°–80° F

If MRT is not equal to dry bulb temperature, the difference is to be offset by changes in the dry bulb temperature of MRT (1° of MRT equals 1$\frac{4}{10}$° of dry bulb temperature).

Relative humidity	20%–60%
Air velocity	10–45 feet per minute (fpm)

0–50 fpm is unnoticed
50–100 fpm is pleasant
100–200 fpm is noticeable but still pleasant
200–300 fpm is annoying, drafty—paper flies from desks
over 300 fpm is uncomfortable and requires corrective measures

---

Forced-air systems promote heat gain and loss from the body by moving air over the surface of the skin (convection). The interior surface temperatures in an earth shelter influence the body by either radiating heat to it or receiving heat from it. Air movement and convective heat loss are minimal. Because of these differences, the same air temperatures in the two buildings can have different effects on the perception of comfort.

Factors Affecting Comfort

Figure 3-4 is a schematic representation of the comfort zone. Ideally, it lies between 30 and 65 percent humidity (horizontal axis) and 68° to 84° F, dry bulb temperature (vertical axis on left). The goal of architectural design is to control environmental factors so that this comfort zone is maintained as much as possible. Climatic variations for regions in the United States

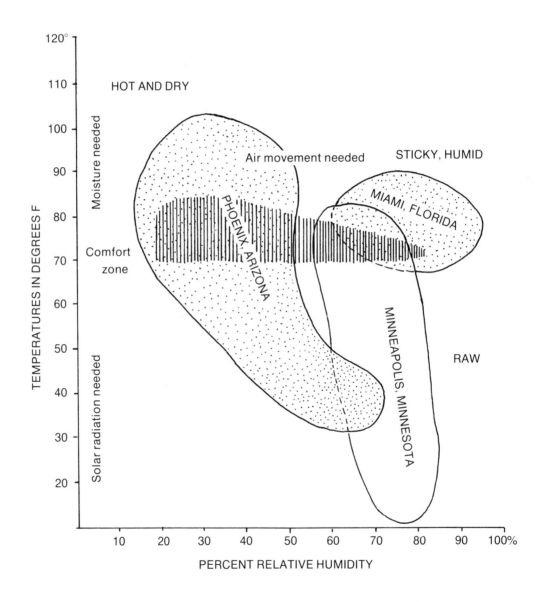

3-4. Climatic chart showing the comfort zone and regional conditions relating to it.

are also indicated on the chart. Some regions are naturally more comfortable for longer periods of time than others.

Our perception of comfort is influenced by changes in air movement, vapor pressure, evaporation, radiation, and amount of clothing. A temperature of 68° F seems comfortable if a light sweater is worn because of the insulative effect of the additional layer. With adequate air movement, temperatures of up to 90° F can feel comfortable. Higher than that, air movement by itself cannot produce comfortable temperatures.

Likewise, air movement and evaporation have limits on their ability to modify the effects of humidity. For example, if an earth shelter is built in the South where interior temperature and humidity cannot be ameliorated by natural means (air movement, earth cooling, or evaporation), some form of mechanical system, such as air conditioning, will be necessary to achieve comfort levels.

Internal surface temperatures of a building may also influence the heat gain or loss to the body and thus affect the sensation of comfort. The MRT (area-and-distance weighted average of the surface temperatures encountered in a space) of a room can be a significant factor in an earth shelter since the earth in contact with the walls, roof, and floor will affect the temperatures of the interior surfaces. Solar radiation collected and stored on these surfaces will also influence the MRT. (Thermal storage walls are valuable in this regard.) Although it is difficult to assign an exact value to the MRT-to-air-temperature relationship, it is generally accepted that a 1° F drop in MRT can counteract a 1°-to-1$\frac{4}{10}$° F rise in air temperature. For example, a room temperature of 80° F and an MRT of 76° F produces a sensation of comfort equivalent to around 78° F. If the air temperature is 63° F and the MRT is 75° F, the comfort level equals an air temperature of approximately 70° F. This means that we can be comfortable at lower or higher air temperatures if the loss or gain of heat from the body can be counteracted by gains or losses associated with the surroundings. If an earth sheltered house is in a region of low hu-

midity, plentiful sunshine, and cool earth temperatures for summer cooling, it can be designed so that the MRT will be very close to comfort levels using only natural means. These conditions may be found in the regions of high earth shelter suitability in figure 3-2.

Climate-Control Strategies

Table 3-4 lists passive climate-control strategies that do not require fossil fuel energy to operate. The passive means of providing heat is to promote solar gain by incorporating either direct gain through windows, indirect gain through thermal storage walls, or isolated gain

TABLE 3-4

Passive Climate-Control Strategies

Goal	Examples
Promote solar gain	Proper orientation and aperture sizing for windows and other openings; more paved surfaces and dark mulches
Reduce solar gain	Proper orientation and use of overhangs, trees, trellises; operable louvers, shutters
Minimize heat loss or gain by conduction or convection	Insulation; earth sheltering; protection by proper building orientation and site placement, by vegetation and land forms; use of air-locked entries, weatherstripping, window insulation such as shutters and shades
Maximize heat loss by conduction	Earth-contact cooling
Maximize heat loss by convection	Enhance natural ventilation by proper building orientation and site placement and by wind channeling (vegetation, land forms, wing walls, overhangs, proper venting)
Maximize heat loss by evaporation	Allow breezes to pass over evaporating moisture; evapotranspiration of plants on earth-covered roof and surroundings
Maximize heat loss by radiation	Expose surfaces to cool night air
Reduce daily temperature swings	Use thermal mass and moderating effects of soil and construction materials
Reduce humidity	Relocate or eliminate humidity-producing activities; promote ventilation, drainage, and solar exposure; pave ground surfaces
Increase humidity	Add vegetative cover; use lower windbreaks and evaporative cooling

through thermosiphon solar collectors. Solar gain is restricted by providing shade or protection with an earth cover. Natural ventilation can be facilitated by correct building orientation and design and use of vegetation. Evaporative cooling (the dissipation of heat by water evaporation) is useful in hot and arid regions. Radiant cooling (the loss of heat by radiation) is better used in areas where heavy mass can be exposed to cool night temperatures. Earth-contact cooling is facilitated by bringing as much building area into contact with cool soil as possible. (Implementing these design strategies will be covered in chapter 4.)

It can now be seen that the success of earth sheltering depends on fitting the building to climate and site. The climate defines the assets and liabilities with which the structure must cope. The goal of a good site layout is to maximize positive features and minimize negative ones. Although many designs can compensate for less-than-optimum site conditions (for instance using clerestories—overhead windows —to collect solar radiation on a north slope), it is best to find a site where environmental conditions approximate the comfort zone.

PRINCIPLES: SITE SELECTION

In selecting a site for an earth sheltered house, an understanding of related social and economic factors becomes as important as awareness of the environmental concerns of climate and microclimate. The social and economic issues associated with earth sheltered housing have already been discussed, but suffice it to say here that issues of location, neighborhood, development pattern, and other constraints can be formidable and should be investigated thoroughly before the final decision to buy a site. Table 3-5 is a checklist of the major considerations relative to site choice that should be examined.

When the general climatic data are known (see table 3-1), it is then possible to evaluate the specific environmental conditions of the site. These concern microclimate (the environment

TABLE 3-5
Site Evaluation Checklist

Physical Characteristics

—geology and soil	soil type and depth, soil fertility, presence of rocks, underlying geology
—water	water bodies, drainage channels, undrained depressions, percolation, water table elevation and fluctuations, seeps and springs, water quality and quantity
—topography	landform patterns, contours, slopes, views, air circulation patterns, accessibility for construction equipment, room for construction activities, erosion problems
—climate	sun angles, precipitation, humidity, cloudiness, wind directions
—microclimates	warm or cool areas; protected areas as shown by plants and snow-melting times; areas of low noise, low odor, and good air quality
—ecology	dominant plant and animal communities, sensitive or endangered species, useful trees and other plants
—manmade structures	buildings, roads, paths, rails, sewers, wells; gas, electric, telephone lines

Cultural Data

—neighbors	number and composition, social attributes and attitudes, current problems
—legal matters	boundary lines, water rights, easements, rights-of-way, access to utilities, zoning restrictions, proposed public improvements, points of ingress and egress

of a small area), topography, orientation, drainage, and soil types. These factors, along with such other variables as roads, power lines, and building code restrictions, can then be integrated with the building plan to produce a truly holistic approach to decision making.

Microclimate, Topography, and Orientation

The components of weather that strongly affect human comfort and thus building performance are temperature, radiation, and wind in a given microclimate. Generally, the coldest temperatures affecting a dwelling occur between 2:00 and 9:00 A.M. and the hottest temperatures between 12:00 and 4:00 P.M. Variations depend on whether the day is clear or cloudy and on the exposure of the site. Clear skies allow significant daytime solar gains, but with the large losses occurring at night, wider temperature variations will be produced than on overcast days. Protected areas are usually prime building sites since the ranges of environmental conditions are more near human comfort levels.

Researchers found that in an Ohio valley climatic data from meteorological stations near airports was significantly different from data recorded at protected microclimatic sites. When compared with the meterological stations, the microclimates had lower summer temperatures, later summer highs, higher winter temperatures, and later winter lows. The protected sites also experienced later frost dates and longer frost-free periods. Furthermore, it has been found that microsites beneath tall grass have half the frost damage than do sites beneath short grass.

Often the most obvious feature of a site's topography is its slope. Contours and slopes may be visually evaluated. Slopes under 4 percent (rising 4 feet in 100 feet of horizontal distance) appear flat, while slopes between 4 and 10 percent appear as easy grades and those over 10 percent seem steep. The latter are prime construction sites for earth sheltered homes (which have been built on slopes of up to 50 percent). The slope of a mowed lawn should be kept under 25 percent. Grades of under 1 percent do not drain well whereas those over 25 percent are subject to erosion, especially where soils are fine grained and subject to intense rains.

It is important to ascertain a slope's stability because certain types of slopes may experience surface sloughing during heavy rains or earthquakes where there are weak zones in the soil. If slope stability is a concern, a professional should be consulted. Features that may indicate potential problems are: hummocks; irregular terrain; abrupt slope changes (portion with low slope angle is generally weak); cracks; stairstep topography; former earthflows; hillside ponds and seeps (look for water-loving plants as indicators); patches of very different vegetation that indicate recent soil disturbances; leaning or canted trees; and planes or joints dipping downslope.

Protected sites on south-facing slopes will experience less severe conditions than open, unprotected areas. South-facing slopes receive much more direct sun in the winter than level sites, resulting in the arrival of spring conditions two to three weeks earlier. Cool air flows downslope and collects in the lower regions of the terrain. Consequently, the top of the slope will be colder than midslope; the valley floor will be very cold. Usually the warm midslope area or thermal belt is the best location for a dwelling because it receives heat during the winter and can be modified to avoid heat gain in the summer.

Because true south should be found when selecting an orientation, the compass must have had the declination adjusted. Declination values can be determined from a map of such values, and most compasses come with instructions for making the rather simple adjustment. If a compass is not available, true south can be determined by simply marking the shadow endpoint of a stake or projection such as a tree at one time and then again after the shadow has moved a few feet. Connect these two points with a line. This line is true east-west in orientation and true north-south is perpendicular to this line.

Wind tends to flow over a hill in a way that causes more turbulence near the hilltop on the windward side than on the lee slope. High-velocity areas are created around the crest with "wind shadows" or "calm-areas" forming near the bottom of the hill. On the crest winds may be 20 percent greater than those on the flats. And, in varied terrain there will be an increase in speed, known as the venturi effect, of winds flowing through a narrow opening. Where the climate is warm enough to require special attention to ventilation, the venturi effect can be used to advantage. Generally speaking, building sites with stagnant, clammy, foggy, frosty, and dusty conditions should be avoided. Such locations usually lie in hollows, below bluffs, or on concave slopes. Also to be avoided are steep north slopes, west slopes facing water, positions at the foot of long open slopes, and bare, dry ground.

Bodies of water act as a thermal drag on land temperatures. Since water warms and cools more slowly than land and air, the proximity of large bodies of water will moderate temperature variations. The presence of water will also raise humidity levels and may affect wind temperature.

The natural cover of an area also significantly affects the microclimate around a home. Temperatures over grass cover can be 14° F cooler than over exposed soil. Areas around marshes, meadows, forests, and other natural communities are influenced even more. Surrounding trees moderate breezes and temperatures and raise humidity levels. Vegetation also improves daylighting by reducing glare; it absorbs dust, pollutants, and many toxic gases (such as ammonia and sulphur dioxide) and reduces noise levels. An earth-covered home actually becomes a part of the microclimate and thus maximizes the benefits of natural cover. For an inland climate, a typical earth-covered roof with vegetation can be expected to dispose of 700 to 1,800 Btu of summer heat per square foot—about 1.5 million Btu per day, per roof top. Trees, however, generally do not belong on earth-covered roofs because they increase the structural loads, especially in windy conditions, and the roots can clog drainage pipes and invade waterproofing systems.

Since vegetation is one of the most beneficial microclimatic assets, every effort should be made to save existing plants during construction. Taking inventory of the plant and animal communities in the area is recommended. On maps of the major plant communities, trees should be identified, and their location, spread, and height recorded. Overlays of the vegetation can be combined with other data, such as topography and soil types, to determine the best place to build. In this way the home will nestle into the site and blend with the natural ecology.

In figure 3-5, the earth sheltered home is an optimum microclimate in the temperate zone.

The gentle, convex, south-facing slope uses the natural currents of air to provide ventilation. A windbreak and hill on the north side protect the house from cold north winds. Oriented to the south-southeast the house receives solar radiation, and promotes natural ventilation by being situated obliquely to the path of summer breezes, a strategy that maximizes the ventilating effects. The natural plant community has been left intact, and the vegetation modifies the microclimate.

Table 3-6 lists the desirable microclimatic adaptations for homes in different regions of the country (also see fig. 3-1). An earth sheltered home in a cool region will perform well on a site that allows solar heating and wind protection in winter. If cooling breezes are allowed to ventilate the house during the hot and humid months, the problems with humidity will also be lessened. Generally, locations that are just below midslope and oriented south-southeast offer the best location for cool regions. In temperate climates, desirable sites tend to be ori-

TABLE 3-6
Earth Shelter Orientations and Slope Positions

Region	Objectives	Orientation	Slope Position
Cool	Retain winter heat Avoid infiltration Receive winter sun Use thermal mass Summer shading Summer ventilation Avoid winter winds Avoid cold pockets	East of south (12 degrees)	Low slope
Temperate	Same as cool	East of south (17½ degrees)	Midslope
Hot and arid	Summer shade Maximize humidity Maximize summer air movement Retain winter heat	East of south (25 degrees)	Low slope
Hot and humid	Avoid summer humidity Summer ventilation Retain winter heat Accept winter sun	South (5 degrees east of south)	High slope

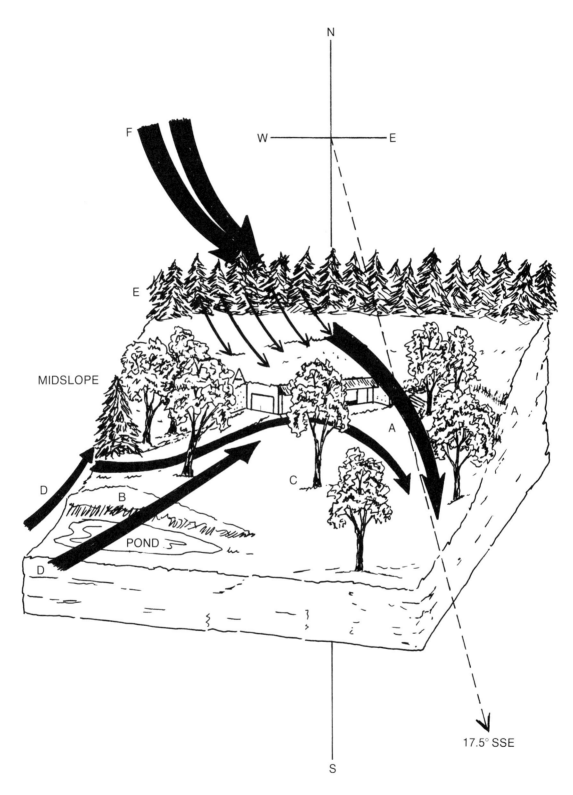

3.5. An earth sheltered home on a site using microclimatic assets. A. Wing walls and berms enfolding house to create air pockets. B. Berm deflecting winds upward. C. Air pocket. D. Summer breezes. E. Windbreak decreasing windspeed. F. Winter winds.

ented farther east of true south and on the middle-to-upper slopes. Lower hillside sites with an east-southeast orientation are preferred in hot, arid zones because of the lesser importance of winds and the large daily temperature range. In hot, humid areas, air movement is critical, and the high air speeds near the crest of a hill are desirable. Both north and south slopes are used in these regions to avoid the excessive summer heat gain associated with east and west slopes.

Often, it is impossible to find the ideal site described above, and a flat area or location that faces in a direction other than south may be the only choice. Occasionally, a breathtaking view may take precedence over energy conservation. If this is the case, the builder should incorporate design features that compensate. Atrium and penetrational designs can be used on certain sites, and adaptations such as clerestories will capture solar radiation and facilitate ventilation.

Each earth shelter building site is unique and must be evaluated individually according to the guidelines given in table 3-5. The factors illustrated in figures 3-5 and 3-10 should also help identify the major considerations. Furthermore, by recording when and where the first snow bank melts or first buds break, and leaves appear, it may be possible to select sites which have subtle advantages that cannot be revealed with physical measurements. Living organisms continually monitor and adapt to environmental factors, often acting as indicators of the conditions on a site. For instance, red maple, alder, tupelo, hemlock, and willow indicate poorly drained, wet ground, while oak and hickory grow on warm, dry land. Spruce and fir inhabit cold, moist places, and pitch pine and scrub oak denote very dry land with good drainage.

Drainage and Soils

Rainfall, drainage, ground water, and soil type are particularly significant to an earth sheltered home because the structure is buried and massive and thus subject to moisture problems and shifting soils. Rainfall directly affects the amount of water in the soil layers near the surface that in turn influences the heat-loss characteristics of an earth sheltered house. Although the effect of rainfall and moisture on thermal performance is largely unknown, it is accepted that moisture reduces the insulating ability of the soil and insulation. Moreover, flowing water can remove heat from the structure.

If possible, low areas, gullies, flood plains, areas with high or perched water tables, and shallow underground streams should be avoided. (The water table is the depth of the main ground water, and a perched water table is a pocket of water located—perched—above the main water table.) This is especially true for atrium-type earth shelters since the water accumulation and drainage problems are more severe than with elevational types. A high water table is indicated by the water levels in wells and diggings, by seeps and springs, mottled soil, and the presence of water-loving plants. A mottled soil points to areas where water has stopped moving through the soil, suggesting a water saturation problem. Flecks of red, orange, and yellow, with possible streaking or patching, will be evidence of water stagnation. Banks, stones, and tree trunks often show the marks of previous flood crests. Throughout the year, both underground and surface water behavior will vary, and the site should be examined in different seasons. If possible, a perforated standpipe driven into the site to an appropriate depth (5 to 20 feet below the foundation level) should be checked regularly to record the fluctuations of the water table.

Monitoring a water well at a site is risky since the water table may fluctuate from place to place, and most wells are driven much deeper than the top level of the water table. A producing well may draw down the water table for many feet in all directions around the site if the soil is permeable. On-site drainage patterns should be mapped so that the future home will not block water run-off routes. The quantity and quality of the potential water supply should also be evaluated by consulting appropriate local agencies such as the Soil Conservation

Service. Laws regulating water rights vary from region to region and should also be researched.

Although not presently required for standard aboveground housing because of the lighter loads, soil testing may be necessary for earth sheltered homes in order to determine such factors as bearing capacity, which affects footing sizes. Foundational design loads for earth sheltered buildings range from 6,000 to 12,000 pounds per foot under the walls supporting an earth-covered roof. These loads, about four times the weight of those of a conventional home with a basement, indicate the importance of investigating the geology and soil types. Years after the structure is built, poor subsoil conditions can cause major problems such as structural cracking and water leaks resulting from uneven settlement or walls that bend and even collapse from expanding soils. Many problems will not show up until years after construction.

Table 3-7 lists the various soil types and their characteristics relative to construction. The soil on a site will probably be a mixture of gravel, sand, silt, and clay and will require a detailed investigation to classify precisely the various combinations. Over 70,000 soil classes have been identified, each with different properties relating to drainage, run-off, erosion, and suitability for construction. Furthermore, since soil types may change within a short horizontal or vertical distance, checks should be made at many points. Also, boulders and bedrock may be irregularly distributed.

Preliminary information about the site can be gathered from the county Soil Conservation Service office (especially if a county soil survey is available), city or county engineer, state and federal geological surveys (see fig. 3-6), previous construction or drilling records, experienced, commercial soil-testing firms, and the local weather bureau. However, as soil conditions often vary across a local region, determining the conditions on the specific building site may be necessary.

Normally, two soil borings are required for a typical house. The resulting information will cover such considerations as location and probable fluctuation of the ground water table; bearing capacity of the soil layers; selection of different types of foundations; and depth of frost penetration. Additional information will include lateral pressures on walls based on both "at rest" and "active" conditions; settlement predictions; percolation, drainage, compaction, and backfilling considerations. Potential problems concerning frost heave, soil corrosivity, or shrink/swell problems with clay are also examined.

If the house is known to bear on continuous shale or bedrock, then a soil test may not be necessary. However, if uncertainty exists regarding the effects of soil type and distribution on bearing capacity, sheer strength of soil, its cohesiveness and shrink-swell potential, a professional should be consulted. It can be disastrous to build a home on more than one type of material. For example, if half the house is on clay and the other half on rock, the foundation will experience differential settling and heaving, seriously damaging the structure and its waterproofing system. More on this later in chapter 4.

It is also possible to make rough identifications of soil in the field during initial surveys of a site by following a procedure described by Kevin Lynch in his book *Site Planning*. If more than half of the particles of a sample of dry soil are identifiable on a clean sheet of paper, it is probably a sand or gravel. If more than half of the identifiable particles are over ¼ inch, it is gravel; if not, it is sand. If less than 10 percent of the total soil sample is fine particles (indistinguishable to the eye), then it is a clean sand or gravel.

Next, remove all the large particles (over ¹⁄₆₄ inch in size), wet the soil, mold it into a pat, and allow it to dry thoroughly. Then take it between the thumbs and forefingers of both hands and try to break it by pressure of the thumbs. If it cannot be broken or is broken only with effort, if it snaps like a crisp cookie and cannot be powdered, then the soil is a plastic clay (CH in table 3-7). If it can be broken and powdered with some effort, it is an organic clay (OH) or a nonplastic clay (ML).

TABLE 3-7
General Characteristics and Typical Bearing Capacities of Soils

Group Symbols	Typical Names	Drainage Characteristic	Frost Heave Potential	Volume Change	Backfill Potential	Typical Bearing Capacity	Range (per square foot)	General Suitability
GW	well-graded gravels and gravel-sand mixtures, little or no fines	excellent	low	low	best	8,000 psf	1,500 psf to 20 tons ft²	good
GP	poorly graded gravels and gravel-sand mixtures, little or no fines	excellent	low	low	excellent	6,000 psf	1,500 psf to 20 tons ft²	good
GM	silty gravels, gravel-sand silt mixtures	good	medium	low	good	4,000 psf	1,500 psf to 20 tons ft²	good
GC	clayey gravels, gravel-sand-clay mixtures	fair	medium	low	good	3,500 psf	1,500 psf to 10 tons ft²	good
SW	well-graded sands and gravelly sands, little or no fines	good	low	low	good	5,000 psf	1,500 psf to 15 tons ft²	good
SP	poorly graded sand and gravelly sands, little or no fines	good	low	low	good	4,000 psf	1,500 psf to 10 tons ft²	good
SM	silty sands, sand-silt mixtures	good	medium	low	fair	3,500 psf	1,500 psf to 5 tons ft²	good
SC	clayey sands, sand-clay mixtures	fair	medium	low	fair	3,000 psf	1,000 psf to 8,000 psf	good
ML	inorganic silts, very fine sands, rock flour, silty or clayey fine sands	fair	high	low	fair	2,000 psf	1,000 psf to 8,000 psf	fair
CL	inorganic clays of low to medium plasticity, gravelly clays, sandy clays, silty clays, lean clays	fair	medium	medium	fair	2,000 psf	500 psf to 5,000 psf	fair
MH	inorganic silts, micaceous or diatomaceous fine sands or silts, elastic silts	poor	high	high	poor	1,500 psf	500 psf to 4,000 psf	poor
CH	inorganic clays of medium to high plasticity	poor	medium	high	bad	1,500 psf	500 psf to 4,000 psf	poor
OL	organic silts and organic silty clays of low plasticity	poor	medium	medium	poor	400 psf or remove	generally remove soil	poor
OH	organic clays of medium to high plasticity	no good	medium	high	no good	remove	—	poor
PT	peat, muck and other highly organic soils	no good	—	high	no good	remove	—	poor

(Source: Underground Space Center, 1982)

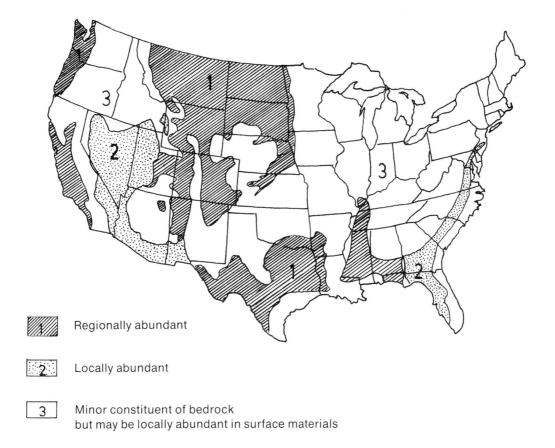

	Regionally abundant
	Locally abundant
3	Minor constituent of bedrock but may be locally abundant in surface materials

3-6. Regions of swelling clay in near-surface rocks in the United States.

The thread test will confirm the above observations. To perform this test, add just enough water so the soil can be molded without sticking to the hands. On a piece of glass (or other nonabsorbing surface) roll it into a thread about ⅛ inch in diameter, then remold it into a ball. Plastic clays (CH) can be remolded like this time after time and will not crack. If the ball cracks when remolded to the thread, it is a nonplastic clay (CL). If it cannot be remolded into a ball once threaded, it is a plastic silt (MH), a plastic, organic silt or clay (OH), or an organic silt (OL). The organic soils (OH/OL) will feel spongy in this test and tend to be darker and have a musty odor if heated when wet. Peat or muck (PT) is identifiable by its black or dark brown color and visible plant remains. In addition, plastic clays have a soapy feel and tend to stain the hands. Between the teeth, sandy soils are hard and gritty whereas silts are not gritty but can still be felt.

In general, gravel bears heavy loads if it is well graded. Sand is good but must be confined at the sides. Fine sands and sand-silt mixtures may become "quick" when saturated, meaning they will flow like a liquid. Silt is unstable when wet and heaves badly when frozen, and peat is a very poor engineering material.

Clay tends to be impervious; when wet, it slips and swells, and it shrinks when dry. The pressures required to restrain swelling can be as much as 2 to 3 tons per square foot. Clay shrinkage can cause unequal settlement of foundations and can open cracks in the soil where water can penetrate. To counter the damaging effects of the shrink-swell cycle, clay soils require special design considerations. These include developing a heavily reinforced structure that resists the forces, placing the building aboveground and berming it, or rejecting the site in favor of another with more desirable soil type. Thick layers of permeable soils on an impervious clay layer on slopes of over 10 percent are liable to slippage. Foundations on fill are to be avoided unless the material is well compacted.

All soil must be checked for its suitability for sewage drainage fields. Gravels and sands, because they are free draining, are recommended not only for drainage fields but also for backfill on side walls. Clay soils are less desirable in this regard. The term *percolation* is used when the dissipation of water into the ground is the primary concern. Percolation rates are measured in minutes per inch of fall of water in a percolation test pit. Sixty minutes per four inches is considered to provide good drainage; per less than 2 inches is poor. A rate of five minutes per inch is fast; a rate of two hundred minutes per inch is slow. Acceptance rates for drainage fields are often around sixty minutes per inch.

The thermal characteristics of a soil can be important to the functioning of an earth shelter because different soil types can affect the thermal lag time and thus the dampening effect of surface temperature extremes. Vegetative cover is also important in this respect and soils that can support good vegetative growth on the roof are important site benefits for an earth-covered structure.

Dr. Sydney Baggs, a landscape architect from the University of New South Wales, has developed a chart that enables designers of earth-covered buildings to estimate the thermal lag properties of various soils. His calculations indicate that the roof of an earth-covered building is best composed of a well-structured clay with good drainage characteristics for subsoil and a composted clay loam for topsoil. If the soil depth on the roof is about 3 feet, this type of soil should provide a lag time of about twenty-five to thirty-five days, that is, surface extremes should take about thirty days to reach the roof of the house. A good vegetative cover on the roof may double this lag time as a thick growth of vegetation may equal the effects of about 3 feet of soil cover in some cases.

Fertile topsoil is necessary for the establishment of living communities for gardens and landscapes. Soil tests that measure drainage, humus content, relative acidity (pH), and the presence of available nutrients (particularly potassium, phosphorous, and nitrogen) will determine fertility. Excess acidity is especially difficult to cope with and should be of special concern. A stable population of earthworms is a good index of the soil pH as they require low acidity to survive.

The pH of the soil may affect the degree to which the concrete and steel of an earth sheltered house corrodes. Gound water, when acidic, can corrode the steel in concrete. A high sulphate content in the soil can also be very detrimental to concrete unless special cements are used, or the concrete is isolated from the soil by waterproofing. Bentonite clays, used for waterproofing, are damaged by high salt concentrations in soil, because salt diminishes the swelling action of the bentonite. Bentonite (montmorillonite) clay is a dry, granular material that forms a highly plastic, impermeable layer when wet (see fig. 5-20).

PRINCIPLES: SITE MODIFICATION

Having evaluated a site in relation to climate, microclimate, drainage, and soils, an owner is now ready to modify the site to enhance the benefits of the proposed home. Sun, wind, and bodies of water are the chief factors in creating a microclimate about a house. Other than building design, the primary controls for sun and wind are shade trees, windbreaks, and earth berms. By wisely locating bodies of water, the humidity, temperature, and reflective qualities of a site can be changed, thereby mollifying climatic extremes significantly.

Sun Control

Figure 3-7 illustrates how the sun crosses the sky in winter, spring, and summer in the continental United States. Notice that the sun is lower on the horizon in winter than in spring or summer. The sun is at its highest point on June 22 and lowest point on December 22. The sun's path across the sky also varies. In the north, the summer sun rises farther to the northeast and sets farther to the northwest. During northern winters, the sun's path is shorter, originating more to the southeast and setting farther southwest. In the extreme north, the sun barely rises above the horizon whereas near the equator it is more directly overhead.

The sun's position is an important factor in passive solar design. In winter (low sun), it is important to place windows, overhangs, and trees so that the house receives the sun's heat. In summer, trees, overhangs, and awnings need to be placed for shading. Table 3-8 gives the sun's azimuths (or bearing angles) and altitudes (heights above the horizon) for different latitudes in the continental United States. By using the table, it is possible to determine the shadow patterns that will fall on a home with various tree placements. Figure 3-8 illustrates the ap-

proximate shading patterns falling on a temperate zone, earth sheltered home during a summer day, with trees located on the west and east sides. (These illustrated patterns are not exact and are for demonstration purposes only.) A cardboard model of the home can be used with an artificial light source held at appropriate angles to determine shading patterns. An overhang, arbor, trellis, or pergola, covered with deciduous vines such as wisteria or clematis, can also shade southern exposures and provide cooling by evaporation.

In cool climates, the southeast to southwest exposures should not obstruct sun in the winter between 9:00 A.M. and 3:00 P.M. when the sun travels from a bearing angle of 150 degrees to 210 degrees and an altitude of 15 degrees in the morning to 20 degrees at noon. Because the winter sun casts long shadows, moderately distant objects that are due south may shade a structure, necessitating their removal. Thick plantings should be placed to the east, north, and west outside of the winter sun's arc. If summer overheating is a problem, a high-crowned deciduous tree close to the south face of the home will provide shade but will not screen off the winter sun. Red maple or eastern sycamore is a good species for this purpose.

In temperate zones, the sun strikes the east side of a home at a low angle throughout the year, transmitting little radiant heat that can be filtered with low density plants such as dogwood or Japanese maple. The south side, on the other hand, receives considerable heat in summer and should be shaded with a high-crowned deciduous tree that will not cut off the low winter sun. Dense foliage, planted away from southerly exposures are ineffective barriers to summer sun and may block the needed winter sun. If they are to be planted for landscaping purposes, careful calculations must be made. Western exposures, which overheat in the summer, should be heavily shaded. A combination of short, low-crowned trees (both deciduous and evergreen) placed on the west and northwest exposures will block the late afternoon sun in summer and filter it in the winter. Apple

TABLE 3-8
Approximate Solar Angles for Most of the Continental United States. (Noon azimuths are always 180 degrees; altitudes at sunrise and sunset are always 0 degrees.)

Latitude	Season	Degrees of Azimuth (Bearing Angle) at Sunrise and Sunset	Altitude Angle at Noon
50 degrees	winter	128	16
	summer	51	63
45 degrees	winter	124	21
	summer	55	68
40 degrees	winter	121	26
	summer	59	73
35 degrees	winter	119	31
	summer	61	78
30 degrees	winter	117	36
	summer	62	83

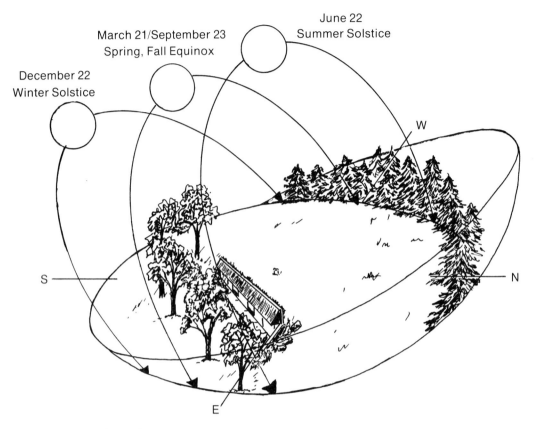

December 22
Winter Solstice

March 21/September 23
Spring, Fall Equinox

June 22
Summer Solstice

S

W

N

E

3-7. Seasonal sun paths in the north temperate zone.

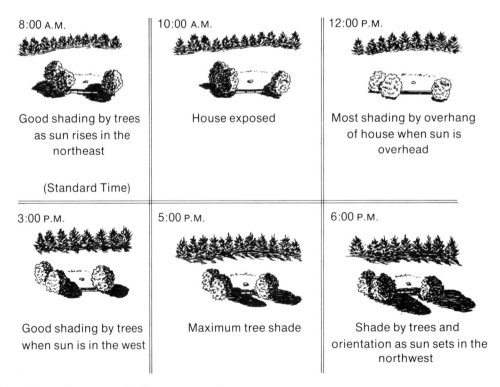

8:00 A.M.	10:00 A.M.	12:00 P.M.
Good shading by trees as sun rises in the northeast	House exposed	Most shading by overhang of house when sun is overhead
(Standard Time)		
3:00 P.M.	5:00 P.M.	6:00 P.M.
Good shading by trees when sun is in the west	Maximum tree shade	Shade by trees and orientation as sun sets in the northwest

3-8. Approximate patterns of summer shading on an earth sheltered home in the north temperate zone.

and pear trees, yews and arborvitae are appropriate species. Medium and tall trees planted at a distance can also provide good shade on the west. Landscaping for sun control on the north is not needed since little direct radiation is received here.

Sun control is also needed on south, east, and west exposures in hot, arid regions. Tall, high-crowned trees such as palm or eucalyptus can be planted on the south and lower growing mesquite or desert willow on the east. On the west, which receives a torrid late afternoon sun, species such as California live oak or fan palms can be planted. Prickly pear and organpipe cactus are effective if planted close to the structure.

In hot, humid climates, complete shade beneath widely spread, high-crowned trees such as palms is desirable. During the morning and evening, distant foliage will provide shade from the low sun. At midday, when the sun is directly overhead, the roof needs protection. Care should be taken not to plant too densely, however, since humidity will build up if cooling breezes are blocked. If trees are not available, vine-covered trellises (using such species as Virginia creeper) can be valuable for shading southern exposures.

Wind Control

Local winds can be controlled with proper landscaping. Air flows like water and can be channeled by the placement of buildings, earth forms, and plants. When a solid barrier blocks the flow of wind, a low-pressure area is created behind the object that suctions wind back into place. In contrast, a pierced barrier allows wind to penetrate, and the low-pressure suction does not develop; this produces less wind reduction but has a greater overall calming effect that extends beyond the barrier. A windbreak of trees acts as a penetrable barrier, effectively slowing the north winds of winter.

Heat loss from a building is proportional to the square of the wind velocity. Wind increases heat loss by convection and by injecting cold air into a building. In cool regions, effective

weatherstripping and a carefully constructed windbreak can save considerable amounts of energy. Figure 3-9 illustrates the effect of windbreaks of different densities on wind reduction. The more penetrable the windbreak, the more widely spread the wind protection will be, although at any one place it may be less than with a dense shelterbelt (a band of trees planted for wind control). Dense coniferous windbreaks placed a distance from the home equal to 3-to-5 times the height of the trees in the shelterbelt will offer the best protection in winter. Protection is lost beyond five times the height of a solid windbreak. If homes must be located far from windbreaks, it is best to use a more penetrable shelterbelt which, though offering more modest protection, does so over a greater area. Irregular windbreaks that have trees of varying heights are most effective. Because a leafless deciduous tree has only 60 percent of the wind-blocking ability of an evergreen tree, a mixture of species and plant sizes provides for the best wind control.

Shelterbelts also affect where snowdrifts form. Downwind drifts near a solid barrier are deep and do not extend much beyond the windbreak itself. Behind a penetrable windbreak, drifts are more shallow and extended, and solid barriers produce drifts on both sides. By careful windbreak design, it is possible to plan where drifts will be deposited.

Breaks in shelterbelts increase wind velocity by releasing high pressure from behind the intact areas of the windbreak (the venturi effect, referred to earlier). For this reason, breaks should be filled unless the venturi effect will be used to provide a wind channel (for example in warm climates when cooling breezes are needed). As a funnel narrows, the wind velocity increases, and a large "scooping" windbreak narrowed to a small opening can increase the velocity of light winds that would otherwise be ineffective for ventilation. Figure 3-5 illustrates how wind protection and wind channeling can be used effectively around an earth shelter. Caution is called for when wind loads on the south wall of an earth shelter might exceed normal, recommended standards. In this case the

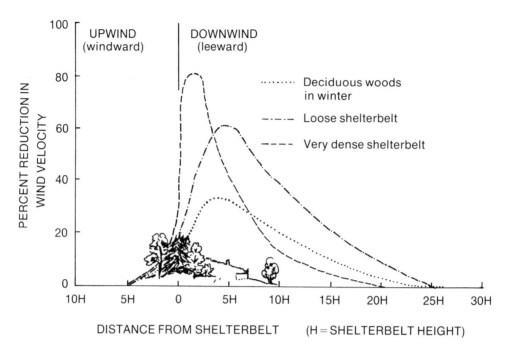

3-9. The effect of a shelterbelt on wind velocity. (Reprinted, with permission from *Earth Shelter Living*, 110 South Greeley Street, Stillwater, Minnesota 55082.)

winds may have to be deflected, slowed, or otherwise channeled away from windows, overhangs, and other structures subject to wind damage.

To control winds, the direction of all air movements on the site must be carefully determined by tying streamers to posts, 5 to 6 feet high, placed at all compass points and in areas of suspected wind pockets. Wind patterns around buildings can also be determined by noting snow deposition patterns in winter and the behavior of smoke and dust in the summer. This information is then used in the placement of trees and shrubs.

Evergreen windbreaks to the south may block solar radiation in the winter unless species of appropriate maximum height are planted. The correct height can be determined by the solar angle for a latitude and its tangent. Most trees should be placed two to six times their mature height from the building to be protected.

Trees, trellises, or overhangs used for summer shading should not block the sun in winter.

Tree foliage should be high enough above the sun-collecting apertures so that branches and trunks do not block the radiant energy. This requires a map of the sun positions for a particular locality (based on altitude and azimuth tables such as table 3-8) and careful siting and pruning of trees and other devices. Even a leafless deciduous tree can block a significant amount of sun and increase heating bills by decreasing solar input.

Fast-growing trees, such as white pine, are appropriate for starting windbreaks but should be replaced by stronger varieties later. A good strategy is to plant one shelterbelt of fast-growing trees and a second of slower, stronger hardwoods within the shadow of the first that it will eventually replace.

In cool climates, thick coniferous trees are planted on the north to block winter winds. In hot, arid zones, home builders must channel winds away from living areas, whereas in hot, humid climes, cool winds that dispel moisture and promote evaporation should be directed into a home. Construction in temperate climates must do both—blocking winter winds and channeling summer breezes. If winter and summer winds come from the same direction, sparse windbreaks planted upwind of the home allow southerly winds to be calmed and will not screen the sun.

Another method of wind control depends on retaining walls and earth berms. Retaining walls (walls at the ends of a house that hold back earth) on earth sheltered homes can funnel the wind stream toward the solar south wall. This may prove advantageous if ventilation is promoted but dangerous if wind loads on windows, patio doors, garage doors, and overhangs are excessive. In regions of high winds and storms, such as the prairie states, it is probably best to create protected pockets by turning wing walls back (see fig. 3-5). Where winds are light, this may not be necessary. A detached earth berm, a mound of earth, which deflects winds up and over the earth shelter may be needed to provide further storm protection (see figure 3-5). However, berms on downhill slopes may catch the cold, descending air, forming pockets of cool air in the winter for which drain areas must be provided. In the summer cool air pockets can be advantageous, and hollows can be used to advantage. The need for detailed analysis of wind and microclimatic factors is underscored by this complexity.

Berms can block low sun, provide wind protection and privacy, and reduce noise. However, berms must be designed with attention to the prevailing winds because a poorly placed one will create unwanted wind turbulence that is especially bad in cool climates.

The Influence of Water

As we have noted, water warms and cools slowly, consuming energy as it evaporates, enabling it to modify the microclimate significantly. The greater the amount of water present, the greater will be the influence, with the maximum effect nearest large lakes and oceans. Temperature differences between water and ground set up convection currents that flow from water to land during the day and the reverse at night. This natural airflow pattern over large bodies of water is a definite aid in ventilation and energy conservation. Small water areas such as ponds do not have this heat storage potential and in cool, temperate, and hot, humid areas should be included more for recreational and aesthetic reasons. However, with improper pond placement, water can reflect unwanted heat and light, produce humidity, and attract insects. Evaporative cooling techniques (from plants, pools, and fountains) are effective in hot, arid climates and can be incorporated into designs.

EXPERIENCES: REGIONAL AND SITE CONSIDERATIONS

As stated earlier, the northern states, the Rocky Mountain region, and Southwest are considered good regions for earth sheltering whereas the southern and southeastern states are thought to be less well suited to making use of the passive features of earth sheltering (see fig. 3-2). Although, earth shelters have been built in almost every state in the union, most of the activity has been located east of the Rocky Mountains and west of the Mississippi River where the need for buildings that mitigate the extremes of climate and provide storm protection is valued.

The vast majority of earth sheltered homes are elevational structures located on south-facing hillsides in rural areas. Usually the building is bermed rather than built below grade. About half of the bermed homes have conventional roofs, and about a quarter of those below grade are without earth on the roof. Poured-in-place concrete is the most popular construction material.

Regional Differences in Earth Shelter Performance

When researchers at Oklahoma State University studied earth shelter owners in the central plains, they found regional differences in energy use. Energy use totals for earth shelter residences in the southern region of the survey

(Arkansas, Oklahoma, and Texas) were well above those in Kansas, Missouri, Colorado, Iowa, and Nebraska. This was due to a decreased dependence on conventional heating and cooling systems for homes in the north where passive solar heating and earth-cooling strategies were more often used than in the southern homes.

Earth shelters have few problems in the heating season; most do well regardless of winter climate if they are well designed. The problems come with summer conditions. Homes where the humidity is high almost always require some form of dehumidification or air conditioning—especially during the first year when concrete is drying and losing water vapor to the inside. This includes homes in the north central, eastern, and northeastern United States as well as homes in the South and Southeast. It appears that humidity can be a problem in all areas, even owners in the Southwest report some difficulty on rare occasions.

In most cases, adequately sized air conditioners (or heat pumps) can solve the comfort problems associated with cooling and humidity, at less cost than that required for conventional homes. An Oklahoma earth shelter reports that a two- to four-hour period of air conditioning will keep the home comfortable (72°–75° F) for around twenty-four hours. This results in bills at least 30 percent less than conventional. Another study, in the Midwest, found that earth coupling can reduce cooling loads and resulting air-conditioning equipment sizes on the order of 50 percent. A home in Louisiana reports that running an air conditioner for thirty to forty minutes will keep the house cool for twelve hours. However, a humidity problem in Illinois was not solved by a large air conditioner because the unit cooled but did not dehumidify because it could not handle enough air. The temperature control thermostat did not respond to the excessive humidity. A series of dehumidifiers was then used but these added some excess heat.

An earth berm homeowner in the Midwest reports that air conditioning was needed for only ten days in a hot summer climate, whereas an earth-bermed home in Arizona used about one-third of the energy for cooling that a conventional home in the same region required. In Massachusetts, an earth sheltered home used only one-fourth as much as nearby conventional homes, but a resident in New Jersey, although reporting satisfactory summer temperatures without air conditioning, noted unacceptably high humidity. An award-winning home in Maryland included a water-heating heat pump to meet nearly a third of its cooling load (which was reduced by the use of earth tubes). In Wisconsin, a home constructed of steel culverts and featured in many publications had no problems with humidity, while other homes in the area needed some dehumidification or air conditioning. In Minnesota and Michigan only a few homes need air conditioning, while Texas and Florida earth sheltered homes all seem to require cooling. None was needed in California, Colorado, North Dakota, Washington, or Montana. Homes in Ohio, Iowa, and Indiana apparently benefit from air conditioning. Missouri and Kansas tend to vary in owner opinion. A Georgia developer routinely supplies his homes with air conditioning and admits that there is no way to get along without it. High humidity has even been a problem in some southwestern homes, where evaporative coolers had been used for short time periods.

Humidity is a concern in most earth sheltered homes in the South and Southeast, central Midwest, East, and Northeast. Soil temperatures in the northern areas keep air or ambient temperatures in the comfort zone, but dehumidification is still needed. In the South, the soil covering lessens temperature variations, but the interior temperatures are too high to be in the comfort zone, and cooling plus dehumidification is necessary. In concrete homes, humidity is more of a problem during the first year than later years (earth shelters made of wood report less of a problem in this respect).

The heating season is much less of a problem for earth sheltered homes. Regardless of climate, all the homes function comparatively well if they are properly insulated or make use of passive solar gains. Reports exist, however, of homes that are difficult to heat because of too much heat loss to cold walls in direct contact with the earth. There are even some complaints of homes being too cold in the summer because of cool, soil-generated interior temperatures. In most instances, however, inadequate design rather than climate was at fault.

Extensive monitoring with soil temperature probes indicates that soil temperatures around buildings differ greatly from predictions on charts of undisturbed soil temperatures. In some cases the actual temperatures registered around 10° higher than expected. This correlates with studies from Washington, D.C., Kansas, Ohio, and Minnesota, where earth-contact coupling (placing house walls in contact with cool earth) was found to be a low-magnitude phenomenon. It is likely that temperature differences of only 5° F exist between the soil and the structure (especially the roof and the walls). A July cooling rate of less than 1 Btu per square foot was found at depths of 4 to 10 feet below a grass-covered surface next to Williamson Hall, an underground bookstore at the University of Minnesota. The amount of cooling by earth contact for a Kansas earth shelter was calculated to be around 5 kwh per day, which was not as large as was expected. The point was made that the major advantage of earth cover was a decreased cooling load rather than earth-contact cooling. In southern areas, earth-contact cooling may be nonexistent. Although this information does not invalidate the strategy of earth-contact cooling (there are reports of very effective cooling performances by cool soils), it does warrant more caution in the design process.

Site-Related Experiences

If a good site can be found, the earth sheltered home allows the maximization of all the benefits of the site. An earth shelter in Maryland had 75 percent of its heating provided by passive solar gain and around 50 percent by passive cooling (shading and earth contact). In New Mexico, an earth sheltered home predicted that 85 percent of its heating would be provided

by passive solar and more than half the cooling by shading, earth contact, and evaporative cooling.

Most of the homes reported in the literature were oriented within the recommended 30 degrees of true south and are on sloping sites. The summer sun from the west has proved to be especially troublesome for those earth shelters improperly oriented, especially if the west wall is exposed or has windows. In winter, homes exposed to winds from the north had higher owner-reported heating bills than protected homes. Earth sheltered homes with proper shading and wind protection were reported to be around 5° to 7° cooler in summer and 4° to 7° warmer in winter than those without.

An owner in Arkansas estimated that his home's heating and cooling performance as well as lighting would have improved significantly if he had oriented his home south instead of west. According to another owner, his southwest-facing home in Illinois would operate better if it faced southeast and could capture more morning sun in winter and less afternoon sun in summer. A north-facing Texas home avoided much of the summer sun, staying about 20° cooler than the outside temperature. However, other Texas homes, facing south, also avoided the summer sun by using tall shade trees and deep earth cover. The soil, even at depths of up to 10 feet, warms up to 80° in some of these southern areas.

Some northern homes, oriented to the north for aesthetic reasons, still perform well in both winter and summer although south-facing homes in the same areas operated more efficiently, especially in winter. The comparison is difficult to make, however, without more scientific data.

Almost no data exists on the effect of landscaping on actual buildings. It is estimated that wind control, solar radiation control, and the placement of vegetation can save up to 30 percent of an aboveground home's total energy requirements for space heating and cooling. Solar radiation is perhaps the major factor since it can affect every portion of a building's design, from heating and cooling to lighting.

The velocity of wind on a building affects air leakage and thermal conductance over the entire skin of the structure. This is especially true of the north and west sides of a building in the Northern Hemisphere. In winter winds can carry heat away from a building and thus increase humidifying loads in the heating season. Since earth shelters are bermed or covered with earth, this effect is probably less than with surface housing and may vary with the amount of roof and wall surface exposed to the elements (often forgotten in the thermal-versus-covered-roof controversy).

Studies conducted with windbreaks indicate that buildings protected by well-designed shelterbelts use 33 percent less fuel in winter than identical buildings in the open. In a Princeton University study of an occupied aboveground residence, air infiltration was reduced from 1.13 air changes per hour (complete air exchanges) before installation of a temporary tree windbreak to 0.66 air changes per hour after installation. This translates to about 3,900 kwh of energy saved on heating costs. Natural plant cover can transform tremendous amounts of energy through the evaporation process. An acre of turf dissipates around 45,000 Btu per day, thus releasing heat. Temperatures around a home can be 10° to 14° F cooler if surrounded by a layer of vegetation. In Davis, California, researchers determined that large trees lowered temperatures around buildings by at least 10° F and reduced loads on air conditioners that were also shaded by trees. Even temperatures on paved surfaces (which can be 25° hotter than unpaved surfaces) are reduced by at least 7° F. Dense trees such as Norway maple block as much as 95 percent of the sun's radiant light, and even leafless trees block 25 percent or more (a fact to remember when placing deciduous trees in front of south-facing exposures). The shade provided by a tree can equal the output of approximately five 10,000-Btu air conditioners.

It is not difficult to find examples of soil-related problems since foundation repairs and damage to buildings from settling soils and expanding clays cost billions of dollars each year

to repair. In 1982 the Underground Space Center's *Earth Sheltered Residential Design Manual* reported the following problems:

- damaged waterproofing and roof settlement (with water ponding) caused by the rapid settlement of inadequately sized foundations for the soil type. The earth on the roof had to be removed, and the roof, columns, and the foundations were jacked and underpinned.

- collapsed retaining wall caused by soft clay slumping against it during a heavy rainstorm. The wall had to be entirely rebuilt and strengthened.

- foundations larger than necessary for the bearing quality of the soil, costing the owner at least several thousand dollars extra. A $500 to $700 soil test would have provided information that could have significantly reduced time and costs.

- an overdesigned shell on an earth sheltered home, approximately twice as expensive as needed, given the relatively good bearing capacity of the soil on the building site.

- sudden appearance of a subterranean spring, unearthed after granite was removed by blasting. Professional help from a moisture control expert was needed to come up with a remedy.

Expansive clays have the potential for serious damage when soil volume changes with fluctuations in moisture content. Figure 3-10 illustrates potential problems that may be encountered in earth sheltered homes not designed for this problem. An earth sheltered home built on a dried clay had the roof raised 4 inches off the walls when the clay beneath the foundations became wet. The forces exerted are formidable. Among the solutions to these problems are: developing a heavily reinforced structure that resists the pressures, maintaining a stable moisture level in the clay around the building, placing the structure aboveground and berming it, or selecting another site. Moisture control methods include the use of over-

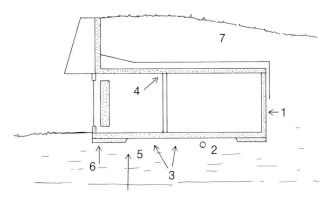

3-10. Sites for potential problems encountered with expansive clay soils: 1. excessive pressure against wall; 2. broken water line; 3. uplift pressures; 4. puncturing of ceiling by column; 5. capillary draw from water table; 6. settling from drying out of soil; 7. poor drainage from roof. (Source: U.S. Department of Energy, 1981)

hangs and gutters to direct water away from the house; sidewalks and other impervious materials to prevent soil drying around the home; and sand layers just below the roof topsoil to prevent water loss by evaporation from deeper soil layers. Berming the structure on grade allows the expansive soils to exert pressure away from the house and permits drainage away from the building.

Firsthand Experiences

Our home is built on a fifteen-acre site (see fig. 3-11), five and one-half miles from town. Surrounded by native grasslands and wheat fields, the site includes a south-facing slope with a drop of about 10 feet in 100 feet (10 percent). There are windbreaks of Osage orange hedge trees on the north and east boundaries and two ponds at the bottom of the slope. The soil is a sandy clay-loam (expansive) over a subsurface base of shale.

Before we purchased the land, I consulted with the Soil Conservation Service and made several site visits with personnel from their office to determine the suitability of the site for earth sheltered construction. In addition to securing a soil report describing soil type and depth based on soil surveys for the local region, I talked with well drillers and neighbors about the nature of the soil and subsurface water. I

also examined the depth of water in nearby wells.

We decided to place the water well on a hilltop away from our planned sewage treatment lagoon. As the well was drilled, I monitored the soil types encountered and recorded water levels. I also secured the driller's soil report. Then, after consulting with soil engineers, I decided that the site was suitable for a heavily reinforced earth shelter.

Although there were no visible signs of slope instability, subsidence, or other dangers, the expansive clay was a problem to be solved. Based on reading the literature about such clays and consulting with soil engineers, we opted for a bermed (but earth-covered) post-tensioned concrete structure with steps taken to control moisture. In a post-tensioned structure, the concrete is poured around sliding cables that can later be pulled tight after the concrete has hardened. Post-tensioning produces a very strong structure in which the floor, walls, and roof are fastened together into a unit. With this in mind, we commissioned a builder of post-tensioned earth shelters, to design our home. We asked him to incorporate certain design features, including provisions for passive solar gain (both direct and indirect) and an elongated, elevational structure that wrapped around the hill.

In selecting the actual location on site, we looked at drainage and microclimate. Finding an unoccupied coyote den on a south slope that appeared to be optimally placed, we made our choice. The drainage was good, and winds were calm here in winter. There also was a channel open to summer breezes from the southeast. True south was determined with a compass (declination was adjusted beforehand), and the building's orientation was set at 20 degrees east of true south to avoid the hot afternoon summer sun. From landscape sketches of the way the house would look in this position, we concluded the site was right. We then outlined the house (with the living room over the old coyote den) on the actual site using stakes and string, and our contractor then outlined the area with lime. Construction began.

3-11. Distant and near views of author's earth sheltered home.

Because of the expansive clay, excavation depth was kept to a minimum of 4 feet and the structure was bermed. A protected pocket was formed by the wing walls and earth embankments (see fig. 3-11), and trees were planted to provide shade and natural ventilation in summer. Sidewalks, placed near the home, provided moisture control of the expansive clays, and the bermed aboveground positioning allowed water to drain away from the house.

We planted natural grasses on the roof, and the native vegetation was preserved as much as possible. This native cover, coupled with windbreaks, sheltered front areas, and the orientation of the house (see fig. 3-5) has created a microclimate that is both energy saving and convenient. Our children can play out front on the coldest days without being subjected to winter wind chill factors. In the summer, the ventilating breezes from the southwest pull air out of the house, drawing cool air in through a short earth tube. This air movement combats high humidity and mildew formation. In summer, however, fans are necessary supplements, and, even so, conditions inside sometimes reach the outer margins of the comfort zone (80° F, 60 percent humidity).

The southeast orientation and curved nature of the house provide real shading benefits, as can be seen in figure 3-12. At our latitude (38 degrees north), the summer sun sets in the northwest, allowing the southeast face of the home to be shaded by about 3:00 P.M. Our home also has a short overhang and movable outside insulation to block additional summer heat. Once the trees are fully grown, the shade will further decrease the heat load. Since providing for summer cooling is of more importance than winter heating, more trees are on the south side. Even with trimming, they will shade some winter sun, but the decrease should be insignificant compared to the summer shading benefits. If necessary, the trees can always be cut or trimmed to provide the proper balance.

Table 3-9 provides the temperature and humidity performance of our home over four years. As the house reaches equilibrium with the soil and the plants begin to mature, the

performance is improving. The addition of draperies and external shutters has had a pronounced effect on both winter and summer performance.

TABLE 3-9

Representative Temperature and Humidity Levels in the Terman Earth Sheltered Home

Year	July Maximum Temperature (degrees Fahrenheit)	Humidity	January Minimum† Temperature (degrees Fahrenheit)	Humidity
1980*	85°	69%	60°	30%
1981*	83°	65%	61°	30%
1982	80°	65%	63°	30%
1983	80°	65%	64°	30%

* No external shutters.
† Vacant house (on vacation).

Soil temperatures around the upper portions of the house are higher than would be expected. In midsummer, the temperature 3 feet down and 3 feet out from the buried edge of the roof in the berm, is about 79° F. Evidently, the soil layers near the roof are more affected by surface conditions than previously expected, and perhaps an earth-covered roof cannot provide significant earth-contact cooling in regions such as Kansas. The soil temperatures beneath the floor have not been measured, but the cooling effects of this area are quite evident, for interior floor temperatures are about 5° to 7° F lower than wall and ceiling temperatures.

Some nonleaking, hairline cracks have shown up in the wing walls, back wall, and floors of the house. No cracks have developed in the ceiling. Although it is difficult to say that these cracks have resulted from expansive soils (as one concrete expert puts it, "Concrete will crack—expect it!"), I believe that the post-tensioning has been important in diminishing potential crack problems. The cracks have not widened over the four years that we have lived in the home. No other problems associated with expansive clays have developed. (See figure 3-10.)

3-12. Morning and afternoon shading patterns in late winter resulting from southeast orientation.

The soil found on-site (a plastic clay-loam) was used for backfilling. Although compacted to a certain extent, the soil still settled approximately 3 inches over a three-year period. It was necessary to add soil and regrade during the first year after backfilling. The soil depth on the roof (3 feet grading back to 2 feet) was sufficient for supporting grass growth, although there was some die-back during the summer (which regrew in the fall). The depth and thermal characteristics of the soil seem to be adequate for providing a thermal lag time of approximately one month and maintaining temperatures around 65° F on the surfaces of the interior walls in winter.

RECOMMENDATIONS

Earth sheltering by itself is rarely sufficient to provide optimum comfort conditions, and in some regions, the use of earth cover might actually make it more difficult to use other, more important climate-control techniques, such as ventilation. Therefore, the design process is of ultimate importance. Earth sheltered homes designed to optimize climatic assets and minimize the liabilities will be successful, whereas those that ignore climatic realities will be less able to provide effective comfort for their occupants.

Given the importance of climate and site, it is probably unwise to purchase plans that are not climate and site sensitive. Such plans may not incorporate positive ventilating currents for every room and may have stagnant corners where mold and mildew will grow. They may fail to provide for the proper amount of glazing for solar applications or may lack the necessary systems to facilitate air handling or conditioning. The building shape may not allow for adequate earth contact (or may have too much), and the insulation scheme may be unable to cope with both winter-heating and summer-cooling needs. These are only a few of the shortcomings that are apt to haunt the uninformed buyer (see chap. 4).

Earth sheltering is not to be viewed as a cure-all for solving all heating and cooling problems but as one strategy, with characteristics that may be advantageous in certain situations and disadvantageous in others. However, used appropriately and designed carefully, earth sheltering can be a powerful force for energy conservation and environmental rehabilitation. Its successful application depends on an informed study of the climatic and site characteristics that can then be integrated with the building design.

CHAPTER 4

DESIGNING THE EARTH SHELTER

*Without consultation, plans are frustrated, but with many counselors
they succeed.*
                                                        Proverbs 15:22

A passive solar, earth sheltered home must be designed with many factors other than energy conservation in mind. If aesthetic and family needs are ignored, the house, although efficient, will prove less enjoyable and fulfilling. Moreover, local building codes must be met, and the design implications for a concrete structure are different from a wood-frame house.

The shape of the building must also be considered; domed structures require different treatment than rectangles, and earth-covered is radically different from earth-bermed. Finally, and perhaps most important, the house must be responsive to the climate and site, described in chapter 3.

PRINCIPLES: THE DESIGN PROCESS

The design process is not a packaged recipe that produces the same results in any situation. Formulating a design requires an attitude toward building that logically evaluates environmental factors and then adapts site and structural components to meet these demands. The design itself represents a sophisticated response to the environment in which owner, designer, building, climate, and site are dynamic parts of an interactive system of resources, needs, and information (see fig. 4-1).

Economics

For most people, cost is perhaps the deciding factor on the type, size, and quality of a future home. Money allows for the mobilization of information and materials into a finished product. As stated in chapter 2 (and detailed in Appendix D), earth shelters can be expected to cost more than aboveground homes. How much more depends on the extent of owner involvement in the planning and building process. Individuals who have construction skills and access to low-priced supplies and labor will be able to build at a cost competitive with surface construction. Others may experience expenditures up to 50 percent more, especially if the site is inaccessible. For some people, this will mean that the project is essentially under the control of an architect or designer and his staff. For others, professional help will be purchased only at selected times (for instance, in structural engineering), and the project will be directed in large part by the owner.

If a professional has prime responsibility for the project, construction, while being more costly, will be less risky since liability for such problems as leaks and structural difficulties will lie not with the owner but with the architect, designer, engineer, contractor, or others hired to do the work. These points must be care-

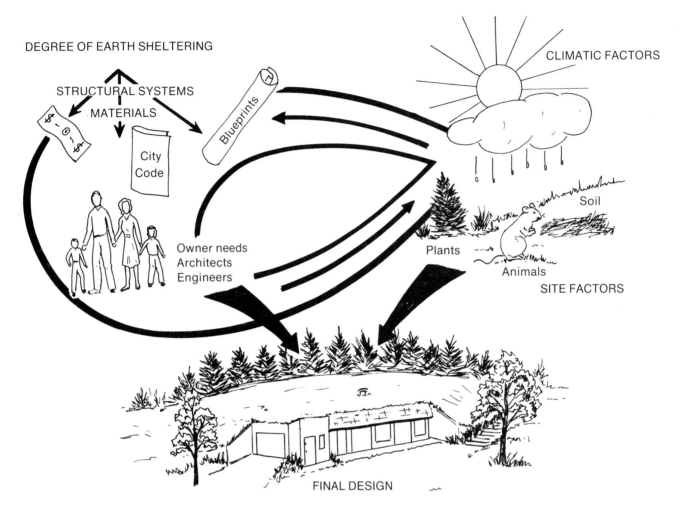

4-1. The design process. Arrows indicate the interaction between the site, climatic factors, and planning.

fully considered in the initial stages of decision making and in the creation of legal agreements (contracts with designers and contractors).

Professional Help

Most owners will need at least some assistance from architectural and engineering services. Furthermore, most banks and loan officials require professional certification of plans, and many building code officials will not issue permits without professional involvement. The challenge then becomes to find and evaluate the quality of professional help.

Not all architects or engineers are familiar with earth sheltering, and the search for qualified personnel can be difficult. Table 4-1 provides a brief checklist for evaluating a potential architect or designer, with the most important criterion being experience (the successful completion of earth sheltered homes that can be investigated and the quality of design determined). (Also see Appendix B for a list of earth shelter personnel.)

Experience and professionalism are also requisite for a contractor. Only those individuals with proven skills in earth sheltered construction techniques should be considered. The

---

TABLE 4-1
A Checklist for Evaluating Earth Shelter Professionals

Professional Training and Involvement
Has the person been trained and licensed?
Is he or she current with the professional literature?

Experience
Has the person been involved in the design and construction of an earth shelter?
Are there examples of their work that can be evaluated?

Legal and financial aspects
Is the person willing to assume liability for mistakes and to indicate this in a written contract?
What is the record of this individual or firm for repairing and paying for any problems (leaks, structural repair, and so forth)?
Is the firm financially able to stand behind their work?

Geographical location
Is the firm or individual located close enough to be available when needed?

Fees and expenses
Are fees competitive with others in the field?
Will the expenses incurred be reasonable?

Design approach
Are the climate and site considered in the design process?

Solar expertise
Is the individual or firm able to make design decisions involving the use of passive and active solar?

Professional relationships
Can the individual or firm show evidence that they can interact with other design professionals and construction personnel?

Personal qualities
Is there evidence of honesty and integrity, such as satisfied customers and solid professional references?
Is there a willingness to explain fully all aspects of the project?

---

contractor must be willing to warranty his work and should readily indicate a willingness to cooperate with other personnel involved with the project. All financial transactions should be carried out in a businesslike manner, and receipts and lien releases should be provided by the general contractor. This is necessary because an unpaid subcontractor has the legal right to bill the homeowner. Therefore, before any payments are made to a general contractor, assurances should be made, mainly in the form of lien releases, that the bills for all materials have been paid.

Degree of Earth Sheltering

The question of how much earth sheltering to use must be resolved early in the design stage. The earth-bermed home is essentially a conventional home with earth against the walls, whereas the earth-covered structure represents a significant departure from the norm and requires a more involved examination.

As previously stated (see chaps. 1 and 2), the decision on whether to build an earth-covered building (soil on the roof) or an earth berm (earth against walls only) depends upon aesthetics, ecology, economics, thermal performance, protection, and maintenance. Figure 4-2 lists the advantages of earth-covered and earth-bermed houses. At present, the major energy benefits of the earth-covered roof involve enhanced abilities to cool in summer, dampen temperature extremes, store solar energy, and moderate heat gains and losses in extreme climates. Although many of these points are disputed, most experts maintain that, everything considered, the earth-covered structure has more potential for energy and environmental benefits than the earth berm (see chap. 1). The risks, however—major construction problems, resale difficulties, and so forth—are greater with the earth-covered home.

As we shall see later, the degree of earth sheltering is a climate-sensitive design response. It becomes obvious, then, that the final decision should be delayed until all aspects of the design process have been considered. Besides selecting the degree of earth sheltering, the soil depth and type of structure (for instance shell versus rectangular) must be considered. This naturally involves questions of materials and structural systems.

Structural Systems

The decision on a structural system (flat roof, domed shell, wood or concrete, and so forth) should be delayed until all factors have been considered. If this decision is made too

Earth-Covered

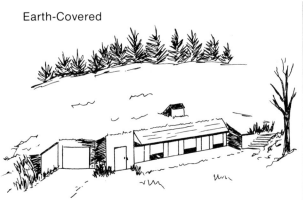

- Reduced infiltration
- Reduced heat loss
- Increased summer heat loss
- Reduced summer heat gain
- Daily temperature fluctuations dampened
- Seasonal ground temperature lag
- More storm protection
- Lower maintenance
- Land use and environmental benefits

Earth-Bermed

- Roof less costly than earth-covered house
- More conventional, hence more marketable than earth-covered house
- Less concern about roof leaks than with earth-covered roof

4-2. Some advantages of earth-covered and earth-bermed structures.

early, many design responses such as adaptation to soil type, ventilation, and solar radiation may be prematurely excluded. For instance, a wood structure may not be best for all soil types and a domed shell may be ill-adapted for receiving adequate solar radiation.

The layout of the interior spaces, however, can be planned at this point and may be a valuable reference in making other decisions.

Auxiliary Spaces, Garages, and Codes

Basements, standard in conventional housing, are too expensive for an earth shelter design. A laundry, furnace, and storage room placed against an earth sheltered north wall is more advisable.

Whether to incorporate an underground garage into the design is a matter of personal preference based on economics and aesthetics and the potential advantages of a mildly heated and cooled space. An aboveground garage is less expensive to build than a reinforced, waterproofed underground garage but is more difficult to integrate aesthetically and lacks the

thermal advantages. An earth-covered garage blends architecturally with the rest of the home and offers passively heated and cooled storage and workspace as well.

Building codes play an important part in the design process. Designed to protect safety, health, and public welfare, they can also be strong determinants in the home's final form.

The major building codes are the Uniform Building Code, Basic Building Code, National Building Code, and Standard Building Code. The housing provisions of all of these have been incorporated into the One- and Two-Family Dwelling Code. A building code becomes law upon adoption by a community or legislature but each community may amend a code as it sees fit. Specifying the requirements that apply in each building situation is difficult. Some rural areas lack any kind of building parameters while other, usually more populated areas, have strictly enforced codes. For this reason, it is necessary to examine the specific code requirements in each community before designing an earth shelter.

The chief areas of concern for earth shelters

are code requirements for fire safety and egress, the effects of grade changes on safety, and provision for light and ventilation. Since these areas were discussed earlier (see chap. 2), this discussion will be limited to suggestions for meeting the stipulations.

Figure 4-3 shows elevational and atrium (courtyard) floor plans that can be expected to meet most egress requirements. Each room has access to a means of escape in the event of fire. Alternatives to this escape requirement (such as doors to separate rooms, separate corridors to fire exits, operable skylights with ladders) must be cleared with code officials. Most officials, however, strongly assert that a bedroom egress window is the best, most direct solution to fire safety problems.

Where grade changes associated with an earth shelter (primarily the edge of the earth-covered roof) present safety hazards to small children, it is advisable to erect guardrails or barriers. This is especially true in urban areas. Alternative design approaches to this problem include preventing access to the roof by fencing or other means, as well as placing a sun screen or trellis below the drop-off to act as a safety net. Unless combined with a guardrail, the use of rooftop vegetation as a barrier is discouraged.

Regarding ventilation, most codes permit the substitution of a mechanical means of introducing air changes that can be incorporated into a forced-air system at little extra cost. Codes recommend that outside air be delivered in specific quantities, based on interior conditions (such as number of occupants, contamination from building materials, and smoke concentrations). Living and bedroom spaces need 10 cubic feet of fresh air per minute (cfm) while kitchens require 100 cfm per room and baths 50 cfm. Figure 4-4 illustrates the use of a mechanical system (heat exchanger) for meeting the code ventilation requirements. Outside air ducted into the heat exchanger is warmed or cooled by the exhausted inside air and is then sent to the living areas. Heat exchangers recover up to 70 percent of heat normally lost by conventional mechanical systems, at a fraction of the cost. To be effective, however, heat ex-

changers must be used in extremely tight homes such as superinsulated types.

Providing natural light for the rooms of an earth shelter is more difficult than for a conventional home. Fewer windows on fewer exposures make it difficult to meet the code requirement for natural light (provided by windows equal to 8 to 10 percent of the floor area) in every room. For rooms away from the windows, skylights, atriums, sunspaces, or clerestories can be used. The codes may also allow for the windows of one room to supply the required light to adjacent rooms, provided the total window area for all habitated rooms exceeds 8 percent of the floor area of all habitable rooms.

Access and Aesthetics

Just as building codes weigh heavily in interior planning, access and aesthetics play major roles in determining the outward utility and appearance of the earth sheltered home. Aside from facilitating solar gain, the site must accommodate construction equipment followed by the normal comings and goings of daily living. Control of visual access from both inside and outside the house is also necessary for protecting privacy and preserving an attractive view.

Aspects of some plans prohibit the use of certain kinds of construction equipment and materials, thereby strongly influencing the final design. For example, if ready-mix concrete trucks are unable to negotiate a grade, then precast panels or wood may have to be substituted for poured concrete. Moreover, if a home oriented to the south presents too little space for parking or other activities, the plan may have to be changed to accommodate clerestories to receive southern radiation. Occasionally, a spectacular view may assume priority over energy-related concerns, and solar may be excluded from the final design. Obviously, access requirements should be examined along with site factors (see chap. 3) at a very early stage.

Unfortunately, in the quest for maximum energy efficiency coupled with structural

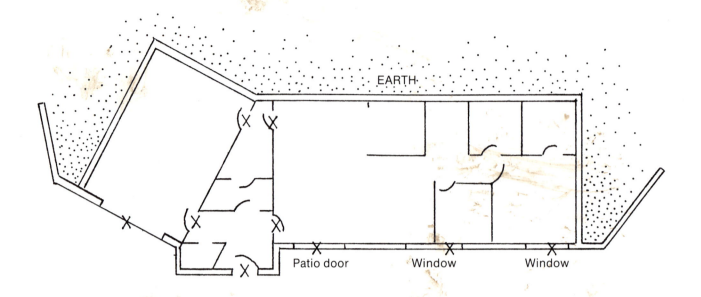

A. ELEVATIONAL FLOOR PLAN

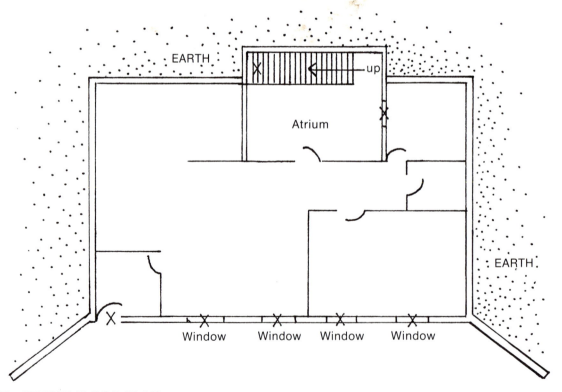

B. ATRIUM FLOOR PLAN

4-3. Floor plans identifying fire escape routes. X indicates exit routes; stippling shows earth in contact with walls. Earth also covers the roof, which is not shown in these diagrams.

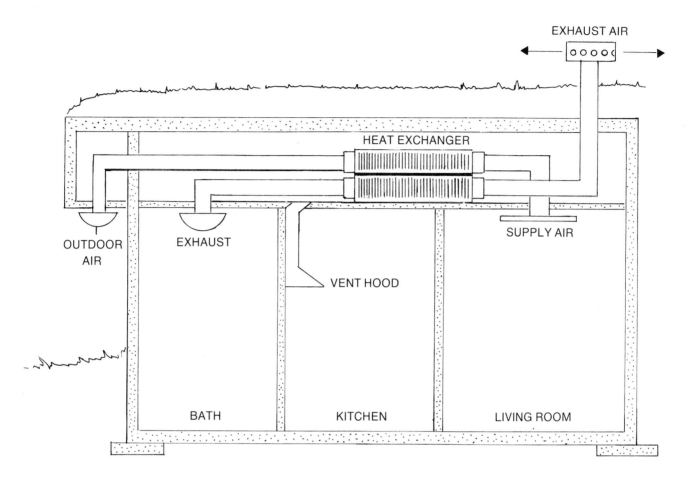

4-4. A mechanical ventilation system for meeting code requirements. Incoming supply air is warmed by the outgoing exhaust of the heat exchange. (Source: U.S. Department of Energy, 1981)

soundness, many earth shelters have been designed that are neither attractive nor appropriate for the site-adaptive philosophy that underlies earth sheltered housing. Because building and site are merged in an earth shelter, it is important to relate the house design to the surrounding character of the natural and manmade environments and unique land forms.

Figure 4-5 illustrates two earth sheltered houses—one in a natural setting and another in a suburban environment. Each home is designed aesthetically to fit its niche. As a general rule, low-profile earth-covered homes enjoy the most potential for blending with natural landscapes whereas the more conventional-appearing high-profile fronts suit manmade

surroundings. Earth-bermed residences also adapt easily to the physical and social surroundings of the suburbs. It should be remembered, however, that aesthetics is only one of the factors in the decision on whether to put soil on the roof.

All of the above elements come readily to mind in the initial design stages. The final decision, however, should be put off until the climatic and site variables are thoroughly evaluated. The cyclic nature of the planning procedures (see fig. 4-1) is a critical realization. One of the contributing factors to poorly designed earth shelters is the inclination to consider only one or two major problems without running through the full course of the design

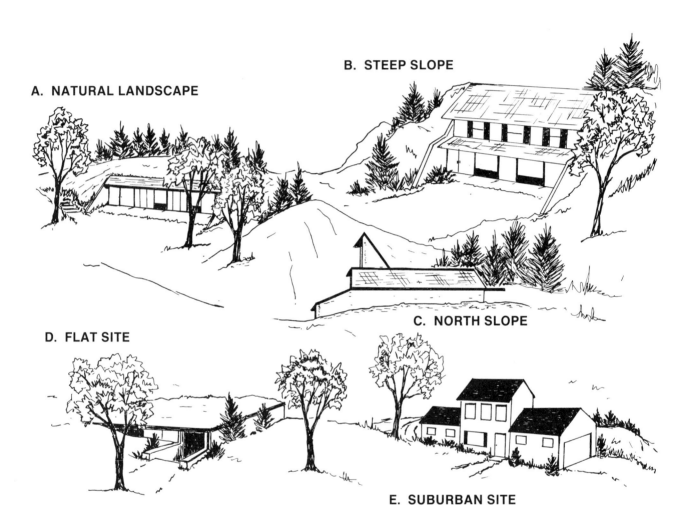

A. NATURAL LANDSCAPE

B. STEEP SLOPE

C. NORTH SLOPE

D. FLAT SITE

E. SUBURBAN SITE

4-5. Adaptations to landscape and topogaphy. A. Natural landscape requires low profile designs that integrate with the surroundings. B. Steep slopes are good sites for two-story designs. C. North slopes need solar exposures to the south, which may be accomplished with overhead windows (clerestories). D. Flat sites may require earth shelters with partially exposed walls to produce an aesthetically pleasing profile. E. Suburban earth shelters may require conventional appearing fronts to blend in with other homes in the neighborhood.

process. For example, structure is such a powerful form generator that often it is "immune" to change when solar accessibility should clearly be given equal consideration. Evaluation, selection, and reevaluation are emphasized, then, in the design's responses to natural environmental components.

PRINCIPLES: DESIGN RESPONSES TO CLIMATIC AND SITE FACTORS

The role of climatic and site factors in determining the potential problems and solutions for earth sheltering was discussed in chapter 3. In this section, the possible responses in building design to these factors are identified and illustrated by an elevational home in a temperate climate. Although conditions vary with climatic regions, the design responses for the temperate regions are the most generalized and applicable. Once these responses are known, each can be evaluated in the context of other design elements (climate, site, and owner preferences), thereby allowing the most adaptive design to evolve.

Table 4-2 lists, in one column, significant climatic and site factors for earth sheltered housing design and, in the other column, the various building systems that respond to each factor. The expected design responses are passive, requiring little or no external energy to operate. Since earth sheltered construction is still an emerging technology, the following discussion certainly is not exhaustive. Many techniques are experimental rather than time tested and should be considered more illustrative than directive. These concepts are to be used as a basis for ideas that may take different forms in different kinds of structures and climates. The adaptive nature of each design response is stressed since the most successful designs will emerge from a careful linkage of specific site factors with appropriate adaptations.

Figure 4-6 illustrates basic cross-sectional designs for earth sheltered homes in cool and temperate, hot and humid, and hot and arid cli-

TABLE 4-2
Environmental Factors and Design Responses

Factor and Aspects	Major Design Responses
Topography (slope direction, slope angle, elevation)	Solar systems, orientation, building shape and floor plan, drainage, waterproofing, structural system, ventilation scheme
Solar Radiation (solar angles, intensity, amounts, terrestrial radiation)	Apertures, thermal mass, exposures, orientation, overhangs and fins, earth cover, shutters, structural system, insulation configuration, interior finishing, mechanical systems, landscaping
Air and Soil Temperature (degree, variations, freezing, thawing)	Same as for Solar Radiation, plus drainage system
Wind and Humidity (frequency, force, direction, abrasive agents, amounts, vapor pressures, vapor pressure deficits, evaporation, transpiration)	Landscaping, orientation, ventilation, shutters, berming, earth cover, mechanical systems, insulation configuration, lifestyle activities, structural system
Precipitation and Soil Water (types, amounts, frequency, snow cover, soil excavation depth)	Earth cover, drainage, waterproofing, structural system, solar system
Soil and Other Loading Factors (profile, structure, texture, moisture, air, clay minerals, pH, other minerals, anions, organic compounds, folding, faulting, weights)	Structural system, landscaping, drainage, earth cover, waterproofing
Fire (combustion)	Materials selection, landscaping
Plants (microclimatic effects, ecology)	Landscaping, earth cover
Animals (potential pests, ecology)	Insulation configuration, earth cover, landscaping
Man (social, legal, amenities)	All systems affected directly but mainly structural system, utilities, floor plan, landscaping

mates. The major responses shown here involve the degree of earth sheltering and grade placement, solar orientation and shading, wind orientation and vent placement, insulation placement and earth-contact exposure. Landscaping and site adaptations are to be used wherever possible (see chap. 3). When these measures cannot provide adequate comfort lev-

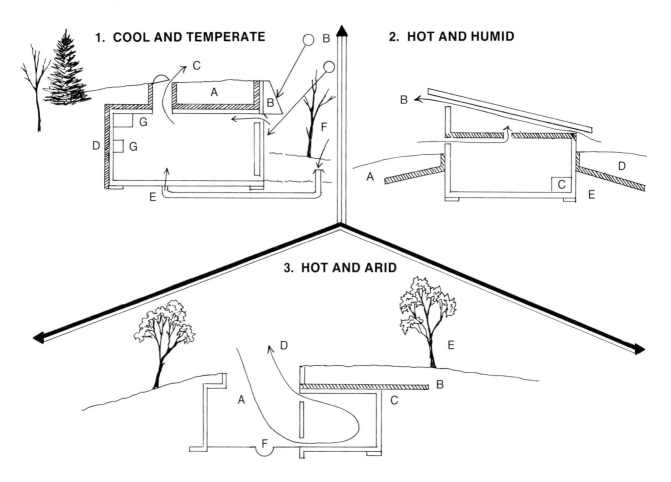

1. COOL AND TEMPERATE

2. HOT AND HUMID

3. HOT AND ARID

4-6. Regional design prototypes for earth sheltered homes. 1. COOL AND TEMPERATE: A. Earth-covered roof. B. Solar orientation and shading. C. Ventilation. D. Insulation. E. Earth contact, earth tube. F. Landscaping, site selection. G. Fans and ductwork for ventilation. 2. HOT AND HUMID: A. Earth-bermed, conventional roof. B. Ventilation. C. Mechanical cooling, venti- lation, dehumidification. D. Insulation that allows earth contact (E). 3. HOT AND ARID: A. Recessed atrium that collects cool air at night. B. Insulation that allows earth contact cooling (C). D. Mechanical ventilation. E. Landscaping, shade. F. Evaporative pool for cooling.

els, mechanical systems may be required. With these basic approaches in mind, it is now possible to consider specific design responses.

Topography

Aesthetically and functionally, the integration of an earth sheltered home with the topography of the land is extremely important. A low-profile home, recessed into a hillside or flat site, will conform more with the natural environment than one with a high profile. Steeply sloping sites require different structural form than gentle slopes or flat ground. Topography also affects sun and wind orientation and site drainage (see below and chap. 3). If a north slope is the only site option, a design that allows for solar input may have to be chosen and fitted to the circumstance. In figure 4-5 each example presents different structural, energy, and other implications relative to the different adaptations involved.

Solar Radiation

Solar radiation provides energy for heating and lighting interior spaces. Passive solar heating integrates logically with earth sheltering that aims to reduce heat loss, provide thermal stability, increase inherent thermal mass, and create adaptable window placement. However, daylighting is problematical because of limited window area, fewer exposures, and the effect of concentrated beams of light from skylights, vents, and so forth. These design challenges must be met in such a way that all solar radiation benefits are maximized.

PASSIVE SOLAR HEATING

Solar angles and the intensity and amount of sunshine varies from region to region. In those areas where the amount of winter radiation compares favorably to the heating demand, passive solar systems should be seriously considered. Where winter cloudiness affects the supply of usable solar energy, it may be advis-

able to stress heat conservation. In general, the adjustment for solar entails increasing glass surfaces (glazing) in areas where a net heat gain can be realized from the sun and decreasing glazing where a net heat loss is likely to occur. The amount of glass is balanced against the need for insulation, and the amount of sunshine makes the difference. Quite expectedly, solar heating is more valuable in sunny, mild regions, and conservation (insulation) is more useful where the winters are cloudy and cold. However, a compromise can usually be reached regardless of the climate, and most earth shelters will benefit from a properly designed passive solar heating system.

Figure 4-7 illustrates the structure and gen-

4-7. Passive solar heating systems for earth shelters. Straight arrows indicate sun rays and wavy arrows, radiating heat. (Source: U.S. Department of Energy, 1981)

A. DIRECT GAIN

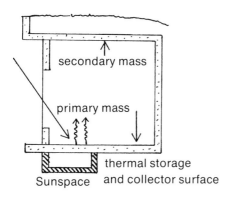

Sunspace · thermal storage and collector surface

B. ISOLATED GAIN

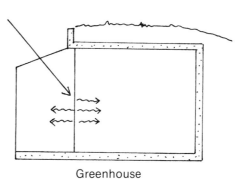

Greenhouse

C. INDIRECT GAIN

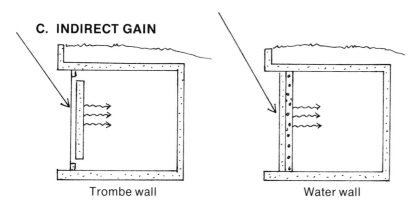

Trombe wall   Water wall

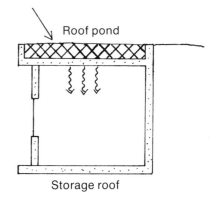

Storage roof

eral function of the most popular types of passive solar systems associated with earth sheltered structures. Each consists of an aperture, collector, thermal storage unit, thermal distribution system, and control components. These elements play critical roles in the functioning of the passive system and require careful integration with the earth sheltered format.

The best earth sheltered format (see fig. 2-2) for passive solar heating is the elevational design in which only the south face of the home is exposed. This format allows for maximum solar gain along an elongated exposure and minimal heat loss because of the extensive insulation and earth cover. The penetrational design is also readily adaptable. The atrium or courtyard plan is more problematical because it lacks a dominant exposure and has built-in shade from the atrium walls. Covering the courtyard with glazing will result in summer overheating and winter heat loss unless measures, which are usually impractical, are taken to cover and shade the glass. Some solar gain is possible, however, and has been detailed by H. Wade in *Building Underground*.

As illustrated in figure 4-7, there are several possible approaches to passive solar heating, through direct gain, indirect gain, or isolated gain systems.

Direct gain systems are essentially well-placed windows for admitting short-wave solar radiation that is then stored in concrete, water, or other material of high thermal capacity. As indicated in figure 4-7, the sun flows through the windows to the living space, which is actually part of a solar collector, where it is stored.

To function properly, direct gain systems require adequate south-facing glazing, often double- or triple-paned and covered with movable exterior shading devices for controlling the amount of heat loss or gain. Furthermore, the correct amount of thermal mass must be available, or large daily temperature swings will occur. This mass must not be covered with interior finishes that may inhibit heat storage and release. Furred-out drywall, panelling, and car-

pet for floors intended for use as thermal mass should be avoided where possible. Isolation of the thermal mass from the outside air and ground is necessary, or too much heat will be lost in the temperature equalization process. An open room arrangement will facilitate the free flow of energy. Where rooms are compartmentalized, mechanical heat distribution may be required.

Indirect gain systems include the mass Trombe wall, water wall, roof pond, and some greenhouses. They are typified by a storage mass located between the living space and the sun that collects and stores solar radiation and then transfers the heat to the living space. The Trombe wall and the water wall readily adapt to an earth shelter. (Roof ponds should be considered for use only in those areas of the country, such as the Southwest, where radiative cooling is pertinent; for all practical purposes, they are rarely used in earth shelters.)

The Trombe wall is usually built of concrete, adobe, stone, or composites of brick, block, and sand. Water walls operate with water as the storage mass. Heat reaches the interior almost immediately or after a delay of up to twelve hours, depending on the depth and time lag property of the material used to construct the wall. As it is with any convective body of water, thermal transfer is rapid within a water wall. This is in contrast to the longer lag time of the concrete Trombe wall. If heat is wanted at certain times, these parameters must be considered in the design. Depending on climate, it is necessary to provide for external shading and nighttime insulation; single, double or triple glazing; selective absorber surfaces (ones that absorb almost all of the solar radiation while reradiating very little); and interior insulation (especially for the water wall, which can cause daytime overheating). Trombe walls that are not suited to climate can lose more heat than they gain.

By incorporating properly sized vents, both the Trombe wall and the water wall can facilitate natural convective air flow. Through openings at the top hot air moves into the living space, drawing cooler room air through lower vents and back into the collector air space. Convective heat distribution is regulated by controllable dampers. A Trombe wall also has the potential to provide for summer ventilation by exhausting hot air externally. Outside air is drawn into and through the living space, thus providing air movement. Another opening must be provided into the living space for replacement air—preferably from a shaded or cooler area (perhaps an earth tube).

With proper design detailing, a Trombe wall might also serve a structural function, thus reducing costs. Water walls almost certainly will add to structural costs.

The third passive solar system, the isolated-gain system, uses a collector-storage component separate from the primary living spaces. Some greenhouses and sunspaces are examples of this type of system, as are thermosiphon, or convective loop systems. Convective-loop systems resemble active solar collectors because they are detached, collect solar heat, and channel this heat to the home. They differ in that they deliver heat by passive rather than active means (pumps, fans, and so on) because the heat flows up to the living space from the thermosiphon collector located below the house. The amount and positioning of the glazing is important and varies with the size and thermal characteristics of the house. Moreover, the amount of thermal mass and method of heat transfer to living space are important design decisions. Isolated-gain systems should have the flexibility to allow for the connection or isolation of the system to the inside of the house, depending on heating needs. Furthermore, shading is necessary to prevent summertime overheating, and vents should be installed to exhaust excess heat and humidity. For earth shelters, the additional humidity of greenhouses is a concern and extra care in planning is required to prevent excessive summer humidity levels.

Some unique designs have been proposed for earth shelters that incorporate double-envelope ideas and vertical crawlspaces. The suc-

cess of these concepts will have to be proven over time, but such activity is witness to the pioneering spirit and dynamic nature of the earth sheltering field.

Selection of an appropriate passive heating system and conservation configuration is a complex decision that may require professional expertise on sizing procedures. Books and articles provide guidelines useful in the initial planning stages (see bibliography), and table 4-3 lists the more popular rules of thumb. Because some of these may be inappropriate for earth shelters, adjustments must be made on a case-by-case basis in order to avoid oversized (and overheated) systems. Usually, an actual, calculated design, relative to heat loss per unit of floor area, is needed, along with accurate solar information for the site. This information is then used, with the help of computers, to arrange and configure apertures, thermal mass, overhangs, and control and distribution systems. A professionally designed system is usually not a great expense when the advantages are weighed against potential owner-designer problems. However, some designers maintain that because of its earth contact and mass, the earth shelter can be adequately designed by using general rules of thumb without resorting to calculations and computer assistance. If this approach is taken, it seems wise to include compensating measures such as movable external shading devices.

Figure 4-8 shows the actual appearance of direct-gain (windows) and indirect-gain (Trombe walls) systems on an elevational earth sheltered home, while figure 4-9 summarizes the total solar design responses of an earth shelter in a temperate climate. Although this home is on a hillside, flat sites with a southerly orientation are also suitable for passive solar (see chap. 3). A variation of 20 degrees east of true south to 20 degrees west will typically change the amount of heat gain by only about 10 percent. However, a southwesterly orientation should be avoided in very hot areas to avoid problems in overheating.

Figure 4-9 demonstrates several principles

4-8. Inside and outside views of window and Trombe wall system.

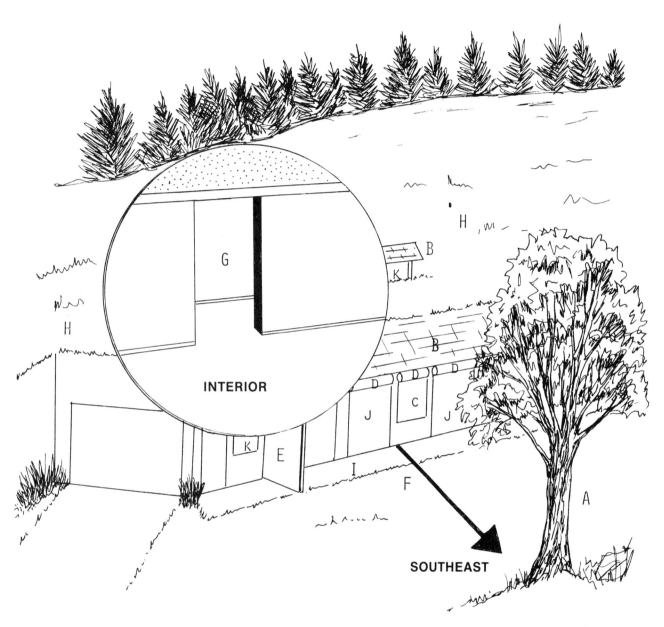

INTERIOR

SOUTHEAST

4-9. Design responses to solar radiation. A. Landscaping. B. Overhangs. C. Windows. D. Movable shading devices. E. Vertical fin for lateral shading. F. Elongated format, earth-covered southeast exposure. G. Interior concrete walls act as thermal mass. H. Earth cover, vegetation. I. Reflective surfaces—sidewalks. J. Trombe walls. K. Daylighting for entrance, walls, and skylight.

TABLE 4-3
Rules of Thumb in the Design of Earth Shelters

Item	Rule of Thumb	Item	Rule of Thumb
Swales	Should course at least 10 ft. from house.	Trombe Wall	Glazing should be 4 to 6 inches from the mass wall; 25 percent of heat from Trombe comes from convection, 75 percent from radiation; thickness of mass wall (concrete) should be about 8 to 15 inches. Trombe wall surfaces should not be over 50 percent of the room's floor area; 300 to 600 Btu per sq. ft. per day heat can come from a Trombe wall; use between .22 and 1.0 sq. ft. of south-facing, double-glazed masonry thermal storage wall for each 1 sq. ft. of space floor area depending on climate and use of night insulation; vent area in Trombes should equal 1 sq. ft. for every 100 sq. ft. of wall area.
Excavation	Standard earth sheltered house will replace 700 cubic yards.		
Weight of Soil	Most soils weigh less than 130 pounds per cubic ft.		
Thermal Lag	Seven days for every foot of soil depth; minimum air temperatures occur 40 to 50 days after 22 December and maximum air temperatures 40 to 50 days after 22 June; at 10 ft. down, the average temperature varies about 5° F.		
Altitude	Air temperatures drop about 5° F for every 1,000-ft. increase in altitude.	Water Wall	About 1 cubic ft. of water for each 1 sq. ft. of solar window.
Atriums	Should be rectangular; should be a drain for every 100 sq. ft.; the angle between the top of the atrium wall and the bottom of the solar glazing should be about 60 degrees minus the latitude.	Sunspaces	Double glaze a sunspace in areas with over 4,000 hdd; 4 sq. ft. of ceiling level heat vent opening is required for every 15 linear feet of house wall separating the sunspace from the room to be heated; sunspace ventilating fans should be large enough to change the air 10 to 15 times every hour; between .33 and 1.5 sq. ft. of south-facing double-paned greenhouse glass for each 1 sq. ft. of building floor area depending on climate; adequate thermal mass should be supplied (.5 to 1.0 cubic ft. of water and 1.5 to 3 cubic ft. of rock for each sq. ft. of south-facing glass).
Solar Orientations	Deviations of 25 degrees east or west of true south will not significantly reduce winter performance; deviations of 15 degrees will influence summer cooling performance, however.		
Thermal Mass	At least 6 times as much heat storage area as direct-gain glazing area; 40 sq. ft. of secondary storage (mass not directly struck by the sun) for every sq. ft. of primary storage (in contact with direct sun); each sq. ft. of sunlight must be diffused over at least 9 sq. ft. of masonry surface.		
Amount of Solar Glazing	Allow .15 sq. ft. of direct-gain solar glass for each square foot of room area; add to this base, 1 square foot of solar glass for each square foot of primary storage and 1 sq. ft. of solar glass for each 10 sq. ft. of secondary storage; .11 to .42 sq. ft. of south-facing glass for each 1 sq. ft. of space floor area depending on climate and use of night insulation.	Air Changes	One exchange with outside air every 1 to 2 hours; use heat exchangers in very tight houses in climates with more than 2,000 heating degree days.
		Natural Ventilation	For daytime ventilation, outside temperatures need to be less than 85° F; for nighttime ventilation outside temperatures need to be less than 70 ° F.
Heat Loss	For every heating degree day, each sq. ft. of floor area in a well-designed earth shelter will require 2 to 6 Btu to keep it comfortable (2 to 6 Btu per sq. ft. per hdd); this equals a heating design load of about 16,250 Btu per hour. Typical gas furnaces supply 90,000 to 135,000 Btu per hour; a person exudes 400 Btu per hour in body heat.	Air Conditioning	1 ton unit for 3-bedroom house; use air conditioning in hot (above 85° F) and humid (above 70 percent) climates where night temperatures are less than 20° cooler than day temperatures.
		Earth Tubes	Use 100 to 700 feet of 4-to-12-inch-diameter pipes buried 8 to 10 ft. deep and sloped to drain; use in climates where summer earth temperatures at pipe depth are below 60° F and humidities below 50 percent.
Clear-day solar gain (40 degrees north lat.)	In January 1,500 Btu per sq. ft. per day though south-facing double-paned glass, 580 Btu per sq. ft. per day in July.	Earth-Contact Cooling	Most effective in climates with summer soil temperature less than 70° F and humidity of less than 30 percent at temperatures above 90° F.
Overhang Projection 40 degrees N	Roughly one-fourth of the height from the bottom of the window to the overhang, for summer shading.	Evaporative Cooling	Use in climates with humidity less than 30 percent.

in action. Although sloped glass receives more solar radiation than vertical glazing, the latter is commonly recommended over sloped apertures on such structures as windows, Trombe walls, and greenhouses in order to avoid the summer overheating, winter heat loss, and water leak problems associated with the sloped approach. Also, vertical skylights are preferred because horizontal skylights are problematical in the control of heat gain and loss. Overhangs and vertical fins correlate with orientation and tree placement (see chap. 3) to provide appropriate shading. Movable external shades allow for adjustments in shading and heat retention. The appropriate amount of thermal mass is exposed to the sun rather than covered with carpet or insulative interior wall coverings, and mass is externally insulated. The amount of earth and vegetative cover is appropriate for decreasing solar gain and for promoting the thermal advantages of earth sheltering (see chaps. 2 and 3). Sidewalks have been placed to facilitate reflective gain.

DAYLIGHTING

Daylighting strategies need to be incorporated with the passive solar features. Earth sheltered homes have fewer exposures and thus more problems in providing adequate natural lighting to meet code requirements. Furthermore, one-directional lighting from a south exposure often causes glare. Entrances into earth sheltered homes should be exposed to natural light to prevent the feeling of entering a tunnel. Light can enter through appropriately placed glazing and overhead openings. By using vegetation and trees and shading devices, glare from south-facing windows can be reduced. White or other pale colors on interior surfaces will compensate for any reduced light by scattering the available light around the room.

Skylights may produce a focused, flashlight-type beam of light. By using clerestory windows and slanting the sides of the skylight shaft, this effect can be moderated. Reflective surfaces on the shaft's sides will also help diffuse light.

Lens and mirror systems, the subject of recent research, may soon be available. Using lenses or mirrors, a beam of light can be directed deep into a building and diffused to provide natural lighting. Good daylighting can cut electrical costs considerably and should be considered in the energy-conservation strategy of every earth sheltered home.

Air and Soil Temperatures

The main design responses to temperature extremes are: control of surface area exposure, choice of insulation and weatherstripping, and control of solar and ventilation systems. Because the interior is isolated from the exterior, these strategies will benefit both heating and cooling efforts.

SURFACE AREA

In order to control heat flow, the amount of building skin exposed to nonbeneficial orientations must be limited. Structural openings on the north experience excessive heat loss, while openings on the west receive excessive heat gain. Table 4-4 presents a computer-generated prediction of the losses in winter heat and summer cooling potential when penetrations are made in the earth cover or insulation skin of different-shaped structures. Obviously, pene-

TABLE 4-4

Computer-Generated Predictions of Winter Heat Loss and Summer Cooling Advantage (Minnesota climate)

Building shape	N, E, or W Penetrations	Winter Heating Requirement (kwh) (October through May)	Summer Cooling Provided by Earth Contact (kwh) (June through August)
Square, elevational	none	4,043	969
Elongated, elevational	none	4,094	822
Square, penetrational	5% on N	4,479	824
Square, penetrational	10% on N	4,916	680
Square, penetrational	5% on W	4,209	726
Square, penetrational	5% on W and E	4,375	523

tration to the north produces the most serious winter heat loss, and double penetration to the east and west most seriously hampers summer cooling.

As mentioned previously, the degree of earth sheltering is important in surface area exposure. As a rule, earth-covered buildings will have less surface area exposed than earth-bermed structures, thereby experiencing less heat loss and gain on a yearly basis. It is usually not economical to place more than 3 feet of soil on the roof unless domed or arched structural systems are used. (The thermal performance of deeply buried shell structures may be considerably better than more lightly covered structures since more of the advantages of soil integration are realized.) Three feet of earth can usually dampen daily temperature fluctuations, an aid in summer cooling. Earth-covered roofs, also important in this respect, provide additional thermal mass, an asset to the passive solar aspect of the house.

The roofs of earth-bermed homes should be heavily insulated (R-value 40 to R-60, depending on climate; see chap. 5) to prevent winter heat losses and summer gains. Although the winter performance of earth berms compares favorably with the earth-covered roof, the summer cooling benefits are less.

INSULATION

Soil is a poor insulator, and heat stored in the structure of the house will be lost unless the home is thermally isolated from the soil. Externally placed insulation is needed to retain heat. For underground areas, water-resistant insulation (which can also resist deterioration from the soil's weight and chemical activity) is needed on the exterior surface of the structure. Exterior insulation allows the thermal mass to store solar energy and thus dampen or delay the heating demands, resulting in lower overall energy requirements, lower peak demands, and more flexibility in the times when heat is supplied. Insulation is critical and its correct installment requires attention be given to such things as R-factors, waterproofing, and thermal breaks.

The amount and positioning of the insulation will vary with the climatic region. In southern homes, insulation coverage will be minimized to promote earth-contact cooling, whereas in cold climates, the retention of heat is more important, and the building shell is wrapped (excluding the floor) with insulation. More details on insulation materials and placement will be given in chapter 5.

The exposed facades of the home should be well insulated, weatherstripped, and caulked, and vapor barriers should be installed. Windows, which have poor resistance to heat flow, should optimally be placed on the south, be double- or triple-glazed and covered with insulating shades or shutters. Skylights should have insulative "plugs" that can be easily inserted and removed to prevent heat loss at night. Movable shading should also be available for skylights to prevent excessive summer heat gain.

AUXILIARY AND MECHANICAL SYSTEMS

An auxiliary or mechanical heating and cooling system for an earth sheltered home is substantially different from that for a conventional home. The mass of an earth sheltered home and its reduced heating and cooling loads decrease the size of heating and cooling systems required. A well-placed woodstove may be all that is needed to provide additional heat in the winter, and a small air conditioner or dehumidifier may suffice for summer comfort.

An oversized heater or air conditioner may not run at the appropriate times to coincide with the heat release of the thermal mass or to be able to dehumidify. Appropriately downsized units will run more frequently but with less energy waste, because the outflow of heated or cooled air is constant and can be evenly distributed. It is also possible to use small, individual room heaters or air conditioners to zone the home. This may obviate the need for extensive duct work and allow more control of individual room temperatures. However, a forced-air system works well in an earth shelter because the duct work can be used to supply heat, distribute cool air, purify air, and promote air movement, thus reducing humidity and air stratification.

(More will be said about heating, cooling, and interior conditions in chaps. 5 and 6.)

Figure 4-10 illustrates the design responses to air and soil temperatures. (See fig. 4-9 for solar aspects.) A design that responds to temperature will emphasize strategies that inhibit the flow of heat into and out of the structure by effectively modifying surface area exposure, earth cover, and insulation, in cooperation with solar control, earth-contact cooling, and auxiliary heating and cooling.

Wind and Humidity

Under winter conditions earth sheltering provides protection from wind by reducing heat loss. However, in summer when air movement is needed, the earth shelter may be difficult to ventilate because of its limited window area. For this reason, it is important to consider the site's wind patterns (see chap. 3), as well as building designs that reduce infiltration in winter and facilitate ventilation in summer.

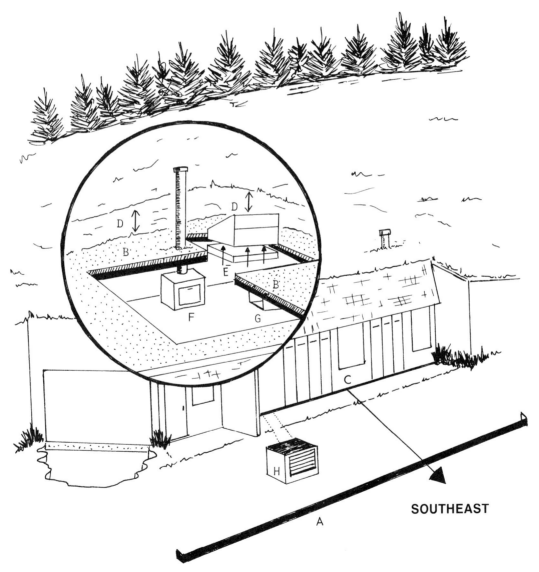

4-10. Design responses to temperature. A. Elongated format; earth-covered southeast exposure. B. Shell completely insulated except floor, which provides earth contact cooling. C. Front insulated and weatherstripped. D. Earth cover of 3 feet; vegetative cover to dampen daily temperature variations. E. Insulative panel for skylight. F. Wood-burning stove. G. Ductwork. H. Earth tube inlet.

ENTRYWAYS

Air-locked entryways are a very important adaptation to wind intrusion in both winter and summer. Reductions in heat loss or gain can be significant if the living space during normal comings and goings can be isolated from outside conditions. A typical air lock consists of a space enclosed by two or more doors, some of which open to the outside and others to the inside.

SUMMER VENTILATION

Figure 4-11 illustrates how a building can provide for summer ventilation. Natural ventilation is achieved by correct building orientation, appropriate site modifications (see chap. 3), strategic location and size of inlets and outlets, and modulation of the interior space. The orientation of the home should allow for the channeling of cool summer breezes that can be captured by vertical projections and overhangs that direct the flow of air into the living areas. The inlets (windows, vents, and so forth) should be small in comparison to outlet sizes. Inlets should be located in high-pressure areas and outlets in low-pressure areas (see fig. 4-11). Outlets can be chimneys, vent pipes from baths and kitchens, or ventilating skylights or clerestories. Both inlets and outlets should have controllable openings so that the house can be sealed when necessary against the outside weather.

The vents associated with a Trombe wall (see fig. 4-8) may also be used in ventilation. Used as a thermal chimney powered by the sun, the openings of the Trombe will admit or exhaust air. A thermal chimney uses the rising hot air produced by solar gain to pull air from interior spaces.

EARTH TUBES

A popular (but largely untested) method of tempering incoming air involves the use of long, underground tubes to warm air in winter or cool it in summer by earth coupling. Air is drawn into the house through 4-to-12-inch-diameter tubes, buried not less than 6 feet. Although it is impossible to make generalizations, most earth tubes are 40 to 400 feet in length, with most cooling occurring in the first 200 feet. A closed loop that starts and ends in the house while bleeding in a small amount of fresh air can also be used.

The major concerns about earth tubes are cost, dehumidification, and possible health hazards. The tubes cannot remove humidity except under ideal conditions. Humid and mild air conditions in the tubes themselves may produce an ideal medium for the growth of molds and microorganisms. The tubes may also be fouled by insects and animals, and therefore caution is advised in their installation. Probably a better option is the heat pump, which is coupled to underground pipes and circulates fluids or air. Such heat pumps are not a totally passive approach, but they do eliminate the above concerns. (More information on indoor air problems will be given in chap. 6.)

Although technically not an earth tube (because of short length and shallow depth), the inlet pipe for the home in figure 4-11 ventilates and provides combustion air to the wood-burning stove. It is also possible to connect an air tube to the perimeter drain around the walls of the home to perform this task.

HUMIDITY CONTROL

As stated earlier, humidity is a significant factor in earth shelter design. Humidity control involves the regulation of humidity input, ventilation, solar exposure, insulation, and mechanical conditioning. Humidity-producing rooms (such as baths and kitchens) should be vented. When outside conditions permit, natural ventilation can be promoted by opening skylights, windows, and vents. Sunlight, which either enters a room or powers a thermal chimney or Trombe wall, will also help to alleviate humidity and air circulation problems.

The problem of condensation can best be appreciated by noting the daily maximum dew point temperatures (see table 3-1) for various regions. Condensation is most likely to occur a few hours after sunset and sunrise. Water vapor

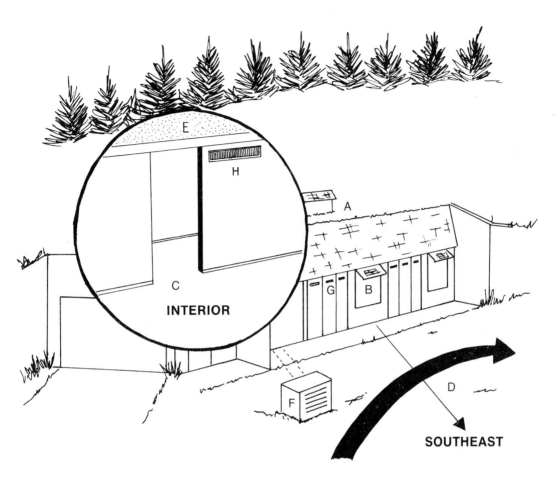

INTERIOR

SOUTHEAST

4-11. Design responses to wind and humidity. A. Ventilating skylight. B. Operable windows, openings, and inlets. C. Open floorplan for air distribution. D. Oblique orientation to the wind (which is channeled by appropriate landscaping). E. Insulated structure to prevent condensation. F. Earth tube inlet. G. Trombe wall vents (shading devices removed). H. Ductwork system for air circulation.

can form upon or behind surfaces such as wood paneling or carpeting that promotes mold and mildew formation.

Insulation is necessary for keeping the temperature of earth-covered walls above the dew-point temperature. If the interior wall surfaces are cool enough to condense water from the interior air, mold and mildew problems will result. Exterior insulation isolates the walls from the soil, allowing them to be influenced more by interior conditions than by ground temperatures. While this is a benefit with regard to condensation, exterior insulation decreases earth cooling. Where wall temperatures are prone to reach dew-point levels, it may be necessary to

reduce humidity levels with appropriately sized air conditioners, heat pumps, or dehumidifiers. An air-circulating system of fans and ducts is also helpful in moving air and reducing humidity.

Precipitation and Soil Water

Leaks are of primary concern to the potential earth shelter owner. With attention to detail, these fears may never be realized, although many earth shelters experience some leaks that must be repaired. To avoid serious and possibly disastrous situations, however, the advice and skills of a professional are usually needed. By understanding the various design responses to water problems, the earth shelter owner is better able to make cooperative decisions with architects, engineers, and other earth shelter personnel.

Precipitation is the cause of most water problems in earth sheltered homes. A 1-inch rainfall can deposit 1,000 gallons or 8,000 pounds of water on 1,600 square feet of roof area (the size of an average roof). This water can enter the home in several ways (see fig. 4-12) and can cause deterioration of structural materials, adhesives, coatings, and finishes. Water (as either vapor or liquid) migrates from areas of high concentration (such as wet soil) to regions of lower concentration (such as building interiors). Water in fluid form can seep through cracks. As a vapor, it can permeate through the pores of concrete, causing excessively high humidity levels inside the house. An effective waterproofing system reduces the pressure of this flow and provides an impermeable barrier between outside and inside.

In areas with frequent downpours and rapid spring thaws, extra attention should focus on surface drainage, backfills, below-grade deflectors (such as polyethylene sheets), waterproofing materials, and subsurface tiles. Figure 4-13 illustrates a generalized scheme that attempts to incorporate these elements. This plan, however, is not meant to be used in all situations because drainage and waterproofing requirements will vary.

DRAINAGE

Ordinarily, homes should not be located in valleys or drainage channels. Instead, a house should be situated where water is drained away in all directions. Backfilling and excavation will help to divert water but should not be depended upon for complete protection. Water diversion channels, (swales) planned into the landscaping design (see fig. 4-14) should be expected to divert water but not completely reroute it. If the latter is necessary, it may be better to select a different site.

Subsurface drainage systems around an earth shelter receive water from adjacent areas and from the roof. Permeable soils and drainage gaps between soil and wall will allow water to penetrate quickly to the subsurface drain tiles. If it is possible that these tiles may be swamped (such as in high rainfall areas), a layer of clay should be placed near the soil surface to prevent excessive water from reaching the pipes too quickly (see fig. 4-13).

In some cases, interior drains may also be designed into the roof to relieve water pressure from flat roofs or decks that have poor gravity drainage. These drains must be designed so that roof movements do not break the connections and cause further leaks.

For water coming off the roof, interceptor drain tiles (or perforated pipes) placed in polyethylene "gutters" along the roof edges (and parapet wall) are recommended. These tiles (covered by gravel and protected from silt clogging by filter mats) will prevent soil saturation and the buildup of excessive hydrostatic pressures, which can force water through cracks and seams into a building. A gravel drainage layer above the final waterproofing layer and beneath the topsoil also helps to promote drainage and reduce water pressures.

For atrium designs, drainage is needed at the roof edges and foundation as well as at the perimeter of the courtyard. If gravity drainage cannot be accommodated, pumping may be required.

To prevent soil saturation and ponding, rooftop and berm areas around an earth shelter should be sloped to channel water away from the house. Swales or tile drains can be located at the foot of slopes to direct water to its final destination. For steep slopes, terracing—a series of steps formed down the slope—is a possibility. All backfill material should be compacted (by one-foot-deep layers) so that soil settling and water ponding may be avoided. Any exposed drains should be screened to keep out rodents or other pests.

Plastic vapor barriers placed below concrete floors will prevent the migration of moisture up through the floor. An underlying porous fill of coarse sandy material is also recommended as is a series of drain pipes that may

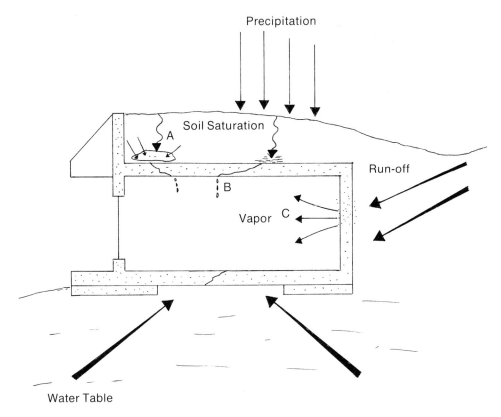

4-12. Some potential water problems in an earth shelter. A. Hydrostatic pressure. B. Capillary draw. C. Vapor pressure raises interior humidity levels.

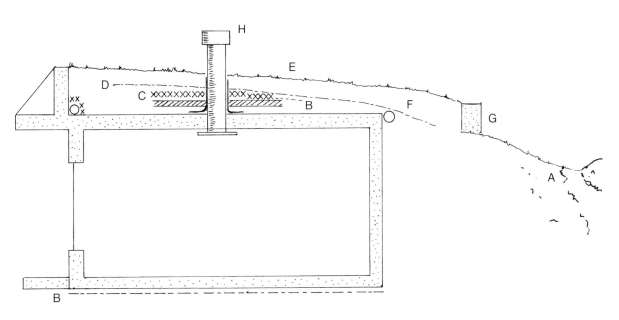

4-13. Design responses to precipitation and soil water. A. Swales (water diversion channels). B. Plastic vapor barriers. C. Gravel drainage lever. D. Clay seam. E. Sloped roof. F. Tile drains (perforated pipes). G. Terracing. H. Roof projections—special detailing required.

occasionally draw down high levels of water. (An exception to this may occur where expansive clays are to be kept wet as a way to control shrinkage and swelling.)

WATERPROOFING

The waterproofing skin is the last barrier between water and the inside of the house. Proper selection and installation are critical. Flat roofs require different materials and techniques than domed roofs, and some materials are better for wet regions than arid (bentonite, for example, a clay-based material, will dry out and crack where it is not kept wet below grade). Waterproofing is an art that depends on the proper selection of materials suited to particular situations. (Additional information on waterproofing materials appears in chap. 5.)

Roof slabs generally should be sloped (between 1 and 8 percent), but not so much that soil slumping, waterproofing displacement, or topsoil erosion occurs. This is especially important in applications that use bentonite. Roofs, unlike walls, must be designed to resist leakage from standing water because moisture will accumulate in pockets or be dammed up behind skylights or other roof penetrations.

Leaks happen most frequently around roof edges and skylights, vents, and chimney flues when slabs and roofs settle or deflect. For this reason, structural integrity is important to waterproofing success, and a reinforcement system, such as post-tensioning, may be a wise waterproofing strategy as well as a structural consideration. Beyond this, extra care is required in waterproofing joints and projections; the number of skylights, vents, and flues should be kept to a minimum; and, according to a recommended strategy, all projections should be grouped into one or two large projections or brought through on the open side of the structure.

Some designers have advocated that the only spaces that should be buried under the earth cover are nonliving areas, such as greenhouses, furnace and utility rooms, and garages, where some leakage may be acceptable. Living and sleeping areas could then be placed under

adjacent, conventional roofs that are buffered by the earth sheltered areas. Most experts, however, maintain that adequate waterproofing can be achieved with proper design.

Soil, Earthquakes, and Other Loading Factors

Figure 4-14 illustrates the loading factors that may influence an earth sheltered home. The loads on the roof can be as much as or more than 600 pounds per square foot depending on the earth cover and building material, while loads against the walls can exceed 120 pounds per square foot, per foot of depth. Besides the physical weights involved, pressures are exerted from water, swelling and heaving soils, slope settlements, and earthquakes. Loads can be permanent or temporary. Soil and structural loads are permanent, but the forces from excavation equipment and rain and snow are intermittent. Obviously, qualified structural engineering, beyond the scope of this book, is required for these situations. Forces from soil factors and other loads are too complex and the consequences too severe to pay less than close attention to something of this importance. Chapter 5 will discuss further details associated with structure, waterproofing, and insulation.

The design responses to loading factors involve the selection and subsequent mode of the structural system (concrete, steel, or wood). Concrete, because of its inherent durability, fire resistance, and high compressive strength, is the choice of most designers. It can be used in various forms: cast-in-place concrete; reinforced or prestressed precast concrete; post-tensioned concrete, reinforced or unreinforced masonry (blocks); or sprayed concrete.

Steel is often incorporated into an earth sheltered structure as part of the roof decking or support system of a concrete building. It is less commonly used for the entire shell of an earth shelter (with the exception of buildings constructed from culverts). Wood, in combination with concrete, is used primarily for roofs, although it has furnished the construction material of the entire shell. However, wood probably

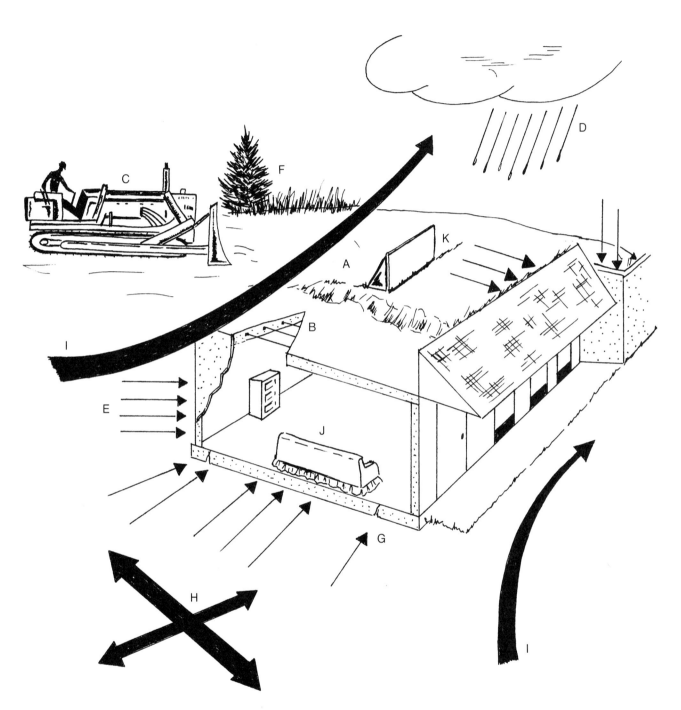

4-14. Load forces on an earth shelter. A. Soil on roof. B. Weight of structure itself. C. Construction equipment and other vehicles. D. Rain and snow load. E. Ground water pressure, swelling clays, slope forces, and settling. F. Vegetation and landscaping loads. G. Frost heave forces. H. Major directional forces of earthquakes. I. Direction of wind forces. J. Interior loads. K. Mechanical equipment on roof (air conditioner, solar collectors, and the like).

will not provide the structural strength required to cope with some of the forces indicated in figure 4-14.

Among loading concerns, the need to tie the walls, floor, and roof together is especially important in regions where earthquakes and tremors may occur. Earthquakes cause heaving and unsettling of the soil around a structure, and square buildings have greater earthquake resistance than elongated buildings with many wings and projections. The waterproofing must be able to reseal any cracks formed during the quake. Wing walls and parapets should be heavily reinforced and anchored, as moving soils will be particularly injurious to these projections. Other potential problems may arise in association with utility hookups, skylights, chimneys, and fireplaces—all of which require special attention in earthquake zones.

Fire

Although fires are liable to start anywhere in an earth-covered structure, they will more than likely begin around the wood-burning stove or in flammable material next to such fire sources as electrical wiring or kitchen appliances. Fires may also be started by lightning or spread from brush or grass fires, although it is unlikely that lightning will damage an earth shelter as much as an aboveground home. Covering a house with earth greatly reduces its exposure to lightning and flying sparks and makes wetting the area down a more manageable operation—another advantage of the earth-covered over earth-bermed home.

In an earth-covered concrete home, structural damage caused by fire should be minimized, but burning interior finishes may produce a hot and dangerous fire with considerable peril from smoke. Because the interior of the home can quickly fill with deadly smoke, a good design calls for well-placed fire exits and judicious selection of interior finishing and exposed fascia materials.

Plants and Animals

Judicious landscaping is part and parcel of the earth shelter design. In chapter 3, the advantages of energy landscaping were detailed. The earth sheltered structure interacts with its biological environment (fig. 4-15). In general, the building's low profile blends with the natural contours of the site. The natural environment is preserved, providing food, shelter, and water for wildlife. The soil cover is adequate for the root growth of plants, and water is nearby to help the plants through periods of drought (which can be severe on rooftop plants).

In order to realize the ecological benefits of earth sheltering, plant selection, soil mixes, drainage requirements, and animal habitats must be carefully studied. Otherwise, animals such as rodents may become pests, damaging insulation and waterproofing systems, or badly chosen plants (such as trees on the roof) can also be harmful.

A certain amount of damage associated with rodents and insects is to be expected, and it is impossible and unnecessary completely to eliminate burrowing animals from an earth-covered roof. Most small rodents will seldom dig deeper than one or two feet, with only minimal damage unless the numbers of animals are high. However, it is recommended that dense fibrous insulation (such as extruded polystyrene, that is, Styrofoam) be used on parapet and wing wall ground-to-surface interfaces. This insulation should be covered with fiberglass or noncorrosive metal flashing down to a depth of 2 feet. Termite protection must be provided where wood is near the ground, although the use of wood at ground interfaces should be avoided.

Although natural areas around the roof will draw most wildlife, some animal invasion will occur during seasonal population dispersal. Such rodent and insect pests can be discouraged by mowing the roof occasionally to reduce cover or filling burrows with bentonite powder to seal them against water intrusion and discourage further use.

Plants also become a problem if undesirable species are allowed to replace the desirable ones and thereby destroy the aesthetics of an earth shelter. Native plants adapted to the region are preferred, but their ecological requirements must be met. Generally, this means providing enough soil depth, correct soil type,

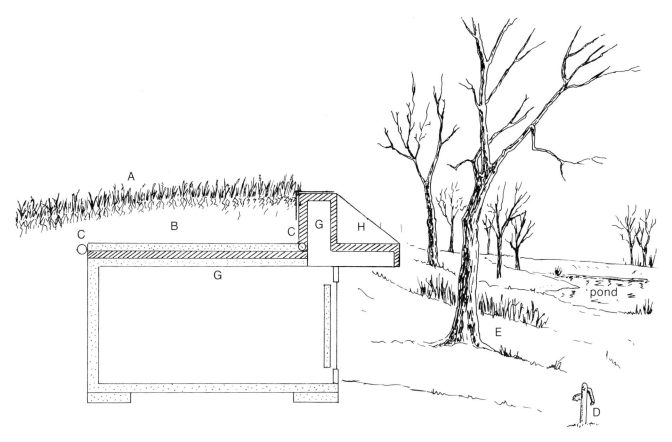

4-15. Design responses to the environment. A. Native plants. B. Fertile soil. C. Surface and subsurface drainage. D. Water supply for irrigation. E. Habitat management; preservation of natural areas. F. Shield against rodents; resistant insulation and waterproofing. G. Structure designed to handle loads. H. Low profile for natural setting.

adequate moisture and drainage, and protection from adverse temperatures and from animals. A good plan is to gather information on the life cycle of the various species so that control measures and fertilizing will be managed at the right times.

Trees on an earth-covered roof can cause structural damage by adding to the weight of the roof cover and transferring a bending force to the roof as the trees are blown by the wind. Tree and shrub roots may also damage insulation and waterproofing systems. Plants grown in containers will solve such problems as well as contribute to the overall landscape scheme.

The Human Factor

The presence and activity of humans probably constitutes the primary design determi-nant for any home. Table 4-5 suggests some of the amenities that may be planned into an earth sheltered residence to make daily living more interesting and convenient. Special areas, such as workrooms, recreation rooms, conversation pits, and so forth, are extremely important to the livability of the home and should be integrated with the home's energy-conserving, structural, heating, and cooling requirements.

Building a small scale model of the home should be one of the final steps before actual construction. The model, made from a commercially available kit or ordinary cardboard, not only helps in evaluating site factors (such as solar shading) and design responses, but also in arranging floor plans.

Chapter 6 will cover the important topic of indoor air quality. Since over 80 percent of a person's time is spent indoors, design adaptations that reduce levels of such pollutants as formaldehyde, nitrogen dioxide, carbon monoxide, and radon are extremely important.

TABLE 4-5
Some Lifestyle Amenities for an Earth Sheltered Home

Porches, patios, and sunspaces

Saunas and hot tubs

Movable carts that fit into spaces at table level

Built-in spaces for television, stereo, telephone, microcomputers, file cabinets, and office supplies

Built-in closets with space for clothes, blankets, shoes, and so forth

Entryway closets, tables, seats, drains, wash areas, and so forth

Storage areas for foods, bulk items, suitcases, and so forth

Workrooms with large worktables, shelves, electrical outlets, lights, and places for tools

Areas for the collection and separation of garbage

Areas for the storage of wood for wood-burning stoves and fireplaces

Rooms for physical exercise and play

Built-in waterbeds

Built-in aquariums (add thermal mass)

Garden work areas (perhaps near greenhouse)

Conversation areas with built-in couches

Shoulder-high walls

Backup hand pumps for water use (in case of electrical failure)

EXPERIENCES: DESIGN EVALUATION AND PLANNING

Both conventional and earth sheltered homes have frequently been built with little regard for the adaptiveness of their designs. As mentioned in chapter 1, this attitude has created much of the energy shortages and environmental difficulties that America now faces. Adaptive design inherently calls for skill, knowledge, and experience, and the resources to put these to productive use. The following accounts are derived from a survey of the literature and visits with earth shelter owners.

Economics and Professional Help

Although money can be saved by avoiding the expense of architects, structural engineers, or solar design professionals, experience reveals that in the long run this omission is usually more costly. Most of the earth shelters in the literature that were built with little or no professional help have been found to be defective in some respect. Cold walls in winter, too little or too much solar gain, condensation or humidity problems, waterproofing difficulties, and structural faults have all been recorded. Problems with code officials have also been documented, and cost overruns up to $95,000 have been reported. Much of the conventional wisdom appropriate for conventional homes is seriously deficient for earth sheltered architecture. Careful selection of architects and engineers and careful attention to details (written agreements, supervision responsibilities, and so forth) are obviously essential.

Many owner-built homes are quite satisfactory in performance, and—more often than not—these owners were wise enough to consult a professional at some stage in the design process. Usually, plans were checked by a structural engineer, and waterproofing and insulation strategies were verified as satisfactory for the particular application.

Occasionally, even professionally designed earth sheltered homes have developed problems due to unknown aspects of earth sheltered design. However, most were correctable (usually at the expense of the architect or other professional involved). Where problems were left to the owner for repair, the professionals hired did not have the financial resources or the inclination to put a warranty on their services. In other instances, the "professional" lacked experience in earth sheltered architecture or was neither licensed nor skilled in the requisite construction techniques.

Table 4-6 summarizes some of the reported problems in earth shelters and the extent of professional involvement. Without professional advice, some serious errors were made. With professional help, most of the problems were less serious, were reparable, and were backed by firms financially committed to their work.

Performance of Design Elements

The success or failure of a design element depends largely on whether it was used in the proper situation. For earth sheltered homes, this means that the designs must fit climate and site characteristics. The following discussion evaluates the performance of elements which were designed with the climate in mind.

DEGREE OF EARTH SHELTERING

Although still a matter for debate, the earth-bermed home is essentially a variant of a conventional home, with increased earth contact. As mentioned earlier, earth berms cost less than (about 10 to 11 percent) and perform as well in winter as earth-covered homes but not as well in summer. This statement is confirmed by reports from the literature and from owners of earth-bermed and earth-covered homes (see chap. 2).

As would be expected, there are more problems with structural integrity and waterproofing in earth-covered roofs, and for the present, the public acceptance of earth berms is greater. Most designers feel that full acceptance of earth-covered roofs by the public will come by the late 1980s.

STRUCTURAL AND WATERPROOFING SYSTEMS

Most of the homes described in the literature are flat roofed and built with concrete, although domed, total wood, and Fiberglas-coated wood earth shelters are becoming more popular. (In fact, the first type of earth sheltered home to be under a national homeowner's warranty program was a total wood structure). Of the flat-roofed concrete homes, many are constructed of precast concrete panels with a topping of poured concrete. Most poured-concrete homes use conventional reinforcement systems, but in recent years, post-tensioning is gaining as the method of choice for many potential earth shelter homeowners. The numbers of domed and barrel-shell (tunnel-shaped) struc-

tures are also increasing. Other earth shelters are built of concrete blocks with precast or wooden roofs.

TABLE 4-6
A Sampling of Reported Earth Sheltered Design Problems

Location	Problem	Remarks
South	No solar gain	Higher heating bills, little professional involvement in design.
	Overheating	Too much heat through skylight, professional was involved.
	Excessive humidity	Poor ventilation, no mechanical systems, some professional involvement.
	Underheating	Inadequate insulation, professional involvement unknown.
South, West	Poor resale	Poor appearance of homes, little professional involvement.
Midwest	Leaks	Bentonite used inappropriately, little professional involvement; poor roof-wall design, little professional involvement; no topping on precast panel roof, professional involvement unknown.
Midwest	Structural damage	Expanding clay, professional involvement unknown.
	Overheating	Atrium covered with glass, little professional involvement.
	Underheating	Cold walls, water drains near walls, professional involvement unknown.
	Structural	Excessive deflection of roofs, incompetent engineering.
	Cost overruns	Problems with subcontractors, some professional involvement.
North	Underheating	Mismatching of furnace with home, some professional involvement; improper placement of heating vents, cold floors—no insulation, some professional involvement.
East	Leaks	Improper use of waterproofing, soaked insulation, some professional involvement; tearing of waterproofing with structural movement, some professional involvement.

As previously stated, barrel shells and domes may have different thermal and air movement characteristics than flat-roof designs. The deeper soil cover of the arched systems allows more thermal lag time and greater interaction with the soil. The higher ceilings and greater interior space may be more difficult to heat and cool, however. Most of the barrel shells in the literature report performances similar to other types of earth shelters with the exception that summer humidity levels may be greater.

The humidity levels in wood (and presumably Fiberglas-coated wood) earth sheltered homes appear to be lower than those in concrete homes, which can be an advantage. However, nonconcrete structures lack the thermal mass distribution and apparently do not store heat and maintain temperatures as well as concrete homes. In regions where solar gain and storage are important, this may be disadvantageous, while other areas such as the South may benefit from a quick response to outside temperatures.

Wood homes may not withstand the loading forces that accompany deep placement or expanding and moving soils, although this is difficult to verify with documented experiences. Leaks cannot be repaired from the inside in wood structures as they might be in a poured-concrete structure.

Cracks in concrete can often be repaired from the inside using a technique called epoxy injection, with which some owners report success. The crack is filled with a strong and resilient epoxy resin that not only provides waterproofing but adds strength. (For more information on epoxy injection, write to Carter-Waters Corporation, P.O. Box 19676, Kansas City, Missouri 64141.) Most earth shelter owners, however, have had to repair leaks from the outside.

Auxiliary Spaces, Garages, and Codes

In the majority of earth shelters, garages are either built underground along with the house or placed on the surface nearby. Earth sheltered garages are reported to be relatively warm in winter and cool in summer and often double as workrooms, laundry, furnace, and storerooms. Surface garages, unless heated, are intended only for vehicular storage. The cost of the earth sheltered garage is significantly more, however.

The professionally designed homes in the literature have floor plans that meet current lifestyle needs and code requirements and are thus attractive to a wide range of people. Many owner-built homes, on the other hand, have unique or custom-built features and room arrangements that do not meet code stipulations and therefore limit their appeal and resale potential.

Access and Aesthetics

The costs associated with construction access have been overlooked by some owners, and the expenses of labor have greatly exceeded expectations. In one case, an earth shelter in a mountainous region cost about 20 to 30 percent more because new roads had to be built for trucks and equipment. In other instances, architects were able to foresee such difficulties, and new sites were chosen or construction techniques were modified to fit the situation.

As suggested earlier, occasionally solar access was sacrificed for a handsome view. Although this involves the loss of solar gain for some houses, others used adaptations such as clerestories that opened to the south while the house windows framed the favored view. These adaptations required almost always the intervention of an architect.

One home, successful in most other respects, neglected to provide space for the parking and maneuvering of cars, thus inconveniencing visitors who had to park at a considerable distance.

Most of the earth shelters detailed in the literature resemble a somewhat conventional ranch house on the exposed side. Architect Malcolm Wells maintains that there remains much to be done to enhance the aesthetics of most earth shelters. At present, the majority of owners, seem to prefer the conventional appearance.

Environmental Factors

Many earth shelters use solar energy through passive solar systems with greenhouses, Trombe and water walls, earth tubes and below-floor storage systems. In many cases, experience has shown that some of these features did not work as well as originally perceived, mainly because of poor matching of climate and system. Other reasons include difficulty in accommodating passive systems with desired views, heat retention, and interior lighting requirements.

Earth tubes, in general, have also performed poorly over a period of time although some successes have been documented. Although temperature reductions of about 5° to 14° F and humidity ratio decreases of about 70 percent have been achieved, earth tube systems have generally been too expensive and the benefits too small for practical uses. Although no diseases involving earth tubes have been reported, it is possible that microorganisms such as those associated with legionnaire's disease could be harbored in earth tubes. Because of these difficulties, many designers are recommending that Trombe walls, greenhouses, water walls, and earth tubes be omitted in favor of direct gain and conventional ventilation systems installed on homes that have properly sized overhangs and are superinsulated.

Below-floor storage systems are well received and have functioned satisfactorily when well insulated and not dependent on moving air through hot rocks or stones for providing heat.

Many owners report satisfaction with passive solar systems. In most cases, these homes are climate-sensitive structures that have been carefully designed by competent solar architects. Well-insulated homes with movable exterior shading devices seem to perform best.

Many homes are not insulated properly, either for heat retention or earth-contact cooling. A lack of insulation may result in cold walls that cannot be heated in the winter, or the wrong type of underground insulation can become waterlogged and lose its insulative prop-

erties. To relieve the latter problem, one owner installed vertical pipes to gain access to buried insulation on the roof. Many homes requiring cooling were insulated so much that the structure was decoupled from the earth and much cooling potential was sacrificed.

A new insulation configuration (developed by John Hait, a Missoula, Montana architect and popular earth shelter writer), isolates a layer of soil between the building and an insulation "umbrella." This scheme has worked well in Montana where a house built by Hait has experienced a yearly interior temperature fluctuation of only 6° F—66° to 72° F.

For heating, many earth shelters rely on a wood-burning stove or fireplace to supply auxiliary heat. Backup heating sources such as heat pumps or gas furnaces are seldom used, although heat pumps appear to work well in earth shelters when supplementary heat is required. Most earth sheltered homes incorporate mechanical equipment (air conditioners or dehumidifiers) for cooling and humidity control, but often these systems are oversized, and the units reportedly do not switch on often enough to dehumidify the home effectively. Only rarely were earth shelters designed and landscaped (see chap. 3) to make use of natural breezes for ventilation.

Most of the homes in the literature have experienced some water leaks caused mainly by structural movement and subsequent damage to the waterproofing. Generally, these have been repaired at moderate expense, although there are some cases of serious design flaws where the problems involved extensive—and expensive —repairs. One involved the omission of a concrete topping layer on precast planks, and another resulted from structural cracks in a home with an inadequately engineered roof and wall union. Because structural movement is a leading cause of difficulties, it follows that those homes with well-designed and secure reinforcement and support systems had fewer problems. Post-tensioned homes are especially well built and have had little structural or waterproofing difficulties.

In earthquake zones, structural damage associated with quakes can be a major consideration in the decision to build an earth shelter. During an earthquake that registered six on the Richter scale, one earth shelter was seriously damaged by lateral movement of the roof and walls, but it did not collapse. The literature also reports earth shelters surviving tornados. In one case, the tornado literally jumped an earth shelter, leaving it undamaged while inflicting severe damage in the surrounding area.

Fire damage to earth shelters is not mentioned in the literature, but many insurance companies offer lower rates to earth sheltered homes. Also, few records of the damage done by insects, rodents, and other pests are available, although serious damage to waterproofing and insulation by gophers occurred in earth-bermed public buildings in Oklahoma.

Examples of Designs

Appendix C provides sketches and brief descriptions of designs considered to be exemplary in the field of earth sheltering. While there are many fine architects, the listed firms direct attention to the use of the climate-sensitive, climate-adaptive approach to earth sheltered architecture. In each instance, an attempt is made to site a home properly. It is also structurally and functionally well engineered to cope with the environment (solar, wind, water, soil, and biological components of the ecosystem). The resulting homes are attractive and very livable.

Firsthand Experiences

Because we wanted a heated and cooled workroom and pantry storage area, we chose an earth sheltered garage. Autos stored in such a garage are not only more comfortable to enter but last longer when protected from the elements. Access (solar, construction, and vehicular) was planned into the site from the start. The home was designed to fit the contour of the hillside in which it was set and to benefit from direct and indirect gain solar systems. Trombe walls were chosen because they could provide a support element as well as solar gain and stor-

age. Movable shading devices (rolling shutters) installed on the outside provided solar control.

Tailored to the Kansas climate, the insulation system retains heat in the winter (thick insulation on roof and thin insulation on the walls) and cools in summer (no insulation on lower walls and under floor). Thermal mass to be used for heat storage was exposed to the sun. A wood-burning stove is the only backup heating source, and no air conditioning was installed. Premolded membranes and bentonite were used for waterproofing, and drainage pipes were installed behind the parapet wall and at the base of the footings. Drainage was further facilitated by placing the home out of any drainage channels and by using long, graded banks to divert water run-off away from the house. All drain pipes emptied by gravity down the slope of the hillside.

The use of such flammable materials as paneling and pressed particle board was kept to a minimum. Every effort was made to maintain wildlife habitat by limiting excavation to as small an area as possible and by replanting with native grasses. The services of the State Fish and Game Commission were also used to provide wildlife food and shelter plantings. Damage from rodents and insects was prevented by covering all exposed wood and insulation with galvanized metal that extended 2 feet below ground. Multicapture mousetraps placed at susceptible entrances (garage, under eaves) are useful especially in the fall, when animals tend to enter houses.

In general, the design has performed well and the home has proved to be very livable as well as an asset to the prairie ecosystem in which it is situated. The solar systems have worked well and supply approximately 85 percent of our heating needs. The direct gain aspects (windows and patio openings) heat the home during the day, and the Trombe walls add to the mean radiant temperature at night. The movable exterior shading devices are very important to the thermal control. Wall temperatures are about 65° F in the winter and 78° F in the summer. Humidity, although a problem in summer, is controllable with fans and natural ventilation. Leaks that have occurred around the vents and skylight were repaired with bentonite from the outside. The post-tensioning system used has resisted the loading forces, and very little cracking has been observed. Some rodents have burrowed into the earth cover on the roof (see fig. 4-16), but none of these has gone below 18 inches (our earth cover is 3 feet), and damage appears to be minimal.

For further information consult Appendix A, which provides data on the energy performance of our home (also see chap. 2). More first-hand experiences with materials will be covered in chap. 5.

4-16. Rodent burrow on earth-covered roof near parapet wall.

RECOMMENDATIONS

Before selecting professional help, it is necessary not only to know the fundamentals of earth sheltering but also to understand climate and site factors that profoundly affect the potential success of a home. In no way should the owners be passive participants in the design of their homes. Homeowners that have construction skills and experience, can direct the building process with only minimal professional consultation. Others will have to rely entirely on the skills of an architect or designer and contractor. In either case, the owners must be able to evaluate professional competence by learning and researching the basics of earth sheltering techniques.

Following is a list of recommendations for beginning the design process.

1. Allow time to become familiar with the field of earth sheltering. Read the available books, attend seminars given by organizations such as the Underground Space Center or Oklahoma State University, and subscribe to magazines such as *Earth Shelter Living*, *New Shelter*, and *Mother Earth News*.

2. Visit earth sheltered homes in your area and talk with owners, contractors, and architects. Note which design elements are working and which are not. Try to ascertain costs and where savings could be made. Get the names and addresses of architects and designers.

3. Decide on the merits of an earth-bermed or earth-covered home. If conventional appearance, low cost, and relatively simple structural and waterproofing concerns are of importance, the earth-bermed home may be the better choice. The need for ventilation (in the South) may also lead to the earth berm. Earth-covered homes are for those people in appropriate environments (see chap. 3) who place priority on maximizing energy efficiency and natural landscape integration.

4. Choose an architect and earth shelter plan with a climate-sensitive orientation. Such items as Trombe walls and other passive systems should be particularly scrutinized with climate in mind.

5. Explore the benefits of the various structural systems. Factors of importance are the availability of materials and services, the accessibility of the site, and the appropriateness of the system for the climate and site fac-

tors. Structural integrity is of the utmost importance in coping with loading factors and water intrusion. Be sure to select a system that is able to withstand the forces on your site, including earthquakes.

6. Plan to make use of passive solar gain and choose the system or hybrid of systems that best fits your site. Remember that vertical glass is recommended over slanted glass on greenhouses and sunspaces; overhangs and movable shading devices should be employed.

7. Insulate the home properly and do not neglect the possibility of earth-contact cooling. Heating and cooling systems must be correctly sized.

8. Pay close attention to matters of access. Choose a site that allows the use of solar and wind benefits and that is readily accessible during construction. Plan where cars and other vehicles will enter and park.

9. Determine the layout of the floor plan by predicting the present and future family needs, including work and playrooms. Build a scale model of the home and move rooms and appliances around to points where maximum efficiency is attained. Remember that earth sheltered garages, although more costly, are valuable in certain situations because of the heated and cooled space available for workrooms and storage as well as vehicular storage.

10. Install furnishings and fascias (fronts) that minimize fire hazard. Also, provide for control of pests and maximize the ecological potential.

Of a good beginning cometh a good end.

John Heywood

CHAPTER 5

BUILDING
THE EARTH SHELTERED HOUSE

And the rain descended, and the floods came, and the winds blew and burst against that house, and yet it did not fall, for it had been founded upon the rock.   *Matthew 7:25*

If properly excavated and poured, framed and waterproofed, insulated, equipped, and back-filled, the earth shelter home will not only look better for longer but may even be more economical to build and live in than homes where little attention is given to construction practices. The most well-thought-out plans and designs are useless unless the appropriate materials and construction techniques are used. When an owner knows what to be aware of in the construction sequence, the right questions can be asked of the construction personnel, thus insuring that all the right steps are taken.

PRINCIPLES: CONSTRUCTION

Because earth sheltered homes are unique, standard building practices are often inappropriate for the demands placed on the structure. Therefore, every stage in the construction process should be carefully scrutinized. Detailed construction information is beyond the scope of this book, but many of the concepts and ideas introduced in this chapter will help prevent major construction failures.

Excavation

As discussed earlier, before excavation begins, the site itself should be carefully analyzed for microclimatic benefits, soil types and their distribution, house placement and orientation, and so on (see chaps. 3 and 4). All points relative to site selection and design must be double and triple checked because the home cannot be reoriented after it is built!

In general, deep excavations are to be avoided. The more earth that must be moved, stockpiled, and replaced, the greater is the probability of encountering rock or water problems—both of which are costly to remedy. Narrow houses that follow the contours of a hillside and that are bermed rather than deeply set are usually less costly and less at risk.

Rock layers usually follow the contours of hills and blasting is often the only alternative where dense rock such as granite and basalt is encountered. Limestone and other sedimentary rock can sometimes be excavated with specialized equipment. An earth sheltered home should be built on rock only when the entire structure can be supported on it. If part of the house sits on separate substrate, differential settling can occur and cause structural problems.

With deep excavations, especially on steep slopes, it is difficult to prevent the up-slope run-off water from damming up against the house and this leads to increased water problems. The water load on an earth sheltered home should come only from the water that falls on the roof. All other water should flow around and away from the structure via swa and drainage channels (see chaps. 3 and 4). Shallow excavations make this easier as is shown in figures 5–1 and 5–2. Furthermore, an earth shelter should not be excavated into the base of a hill or the bottom of a valley because of the increased probability of water drainage problems.

The dangers of hill shift are also increased with deep excavations. On certain slopes the soil may move en masse, damaging structures that are excavated too deeply. Cutting into such a slope for house construction weakens the slope's stability. As this may not occur until several excavations have been made, slope stabilization precautions should be exercised where numerous excavations are planned into steep slopes. Such precautions include limiting excavation to 8 feet or less and berming the

walls of the earth shelter rather than abutting them to excavated hillsides. Figure 5-3 illustrates a hillside after excavation. The depth of excavation is 8 feet deep at one end and only 4 feet deep at the other because the floor plan follows the contour of the hillside. In figure 5-2 the completed home has berms to accommodate drainage channels around the house.

Berms of earth shelters should be contained within property lines. Care must be exercised not to disturb the footings of adjacent buildings in the excavation process. Also, water must not be diverted onto or away from an adjacent site so as to create problems for adjacent homeowners.

The type of excavation equipment will vary with the site conditions (soil type and rock), but common choices are the front end loader and

5-1. Excavation depth and associated problems.

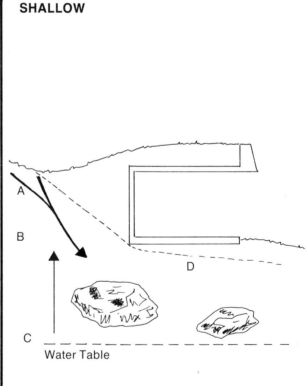

DEEP

Water Table

A. ᴸamming up against house
    d danger of hillshift
    ᴵ soil-water load
        substrate leads to differential settling

SHALLOW

Water Table

A. Water drains away from the house
B. Soil more stable than with deep excavation
C. Water table does not impact structure
D. Uniform substrate

5-2. Rear view of earth shelter showing drainage channels.

5-3. Hillside site for earth shelter after excavation.

backhoe (fig. 5-4). Determining factors may be the size of bucket and the maneuverability of the machine. The backhoe can be used to dig the first sides of the excavation and the loader to remove and pile the bulk of the soil. A small front end loader can be used to backfill and move materials around the site. Operators vary in their ability to run these machines so inquiries are certainly in order before hiring someone. If good operators are available, it is usually better to pay on an hourly basis rather than by the job.

Care must be taken to provide enough room for the operation of equipment. For instance, if a backhoe is to be used to dig footings along the back wall of the excavation, the original size of the excavated area must be wide enough to allow for this. And the excavation must also provide access for cement trucks, concrete pumps, cranes, and other equipment that may be used later (see fig. 5-5).

When the topsoil is removed, it should be piled separately since it will be put on the roof as the seed bed for vegetative cover. Subsoil

5-4. Front end loader, backhoe, and small loader.

5-5. Site must be accessible for construction equipment.

should be kept separate from the topsoil and backfilled or graded first. More on backfilling later.

During excavation, good transit work is critical. The excavation should be planned and carefully situated to provide good drainage and level foundation beds. Rain accumulating in an excavation can slow construction considerably, and a drainage ditch to channel water away from the excavation is recommended. Foundations should be set on solid, compact substrate rather than loose, disturbed soil. Unless care is exercised here, the whole house will settle unevenly.

Ample workspace should be allowed between the future house walls and the banks of the excavation (fig. 5-6). As a considerable amount of work centers around the walls, at least 4 feet of space should be allowed. To prevent cave-ins, the excavation banks may need to be beveled back.

Structural Systems

As discussed earlier, concrete seems to be the preferred material in the construction of earth shelters. It is strong, easily waterproofed, readily available, and construction personnel are experienced in its use. The use of concrete in an earth shelter, requires more accuracy, however, than in a conventional building. The consequences of an oversight are serious because of the significant loads on an earth shelter and the requirements for strength and near-perfect waterproofing.

Also as stated earlier, all plans for an earth shelter should be professionally engineered and then followed exactly. The temptation to deviate from these plans may be strong, but the conscientious owner should not listen to "conventional" wisdom. The differences between an earth sheltered home and a typical basement are considerable and shortcuts acceptable to one are not allowable for the other.

Chapter 4 indicated how concrete can be used in many different forms—reinforced, poured concrete; prestressed panels; post-tensioned construction; sprayed concrete shells; concrete blocks—and in many combinations. In this section, special emphasis will be given to poured, reinforced concrete since more of the construction phases with this form are within the control of the owner and on-site workers. Precast concrete and its placement are generally more uniform and under more rigid control of the supplier and contractor.

The Nature of Poured Concrete

Table 5-1 lists the principal requirements of good concrete and some of the problems that can occur in mixing and installing. The concrete wall in figure 5-7 illustrates these characteristics, most of which concern concrete's three principal components (cement, aggregate, and water) and the way in which the concrete is placed and handled at the job site. The concrete in this wall is well cured and consists of well-graded aggregate that is evenly distributed.

5-6. Workspace between excavation bank and building walls.

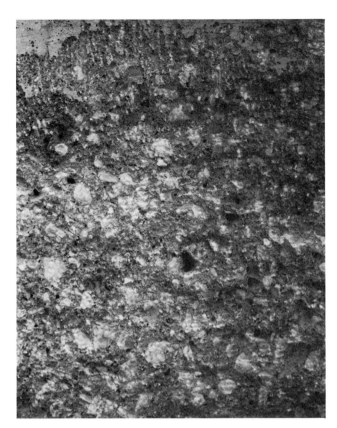

5-7. Concrete wall showing cement and aggregate distribution.

TABLE 5.1
Principal Requirements and Related Problems of Concrete

Requirements	Problems
Water. Must be clean and in specific ratio to cement.	Excessive water decreases strength and durability and causes cracking; excessively rich cement may develop more shrinkage cracks and volume changes
Aggregate. Must be clean and well-graded.	Impurities such as clay, loam, or organic matter cause unfavorable reactions leading to partial or complete disintegration; flat or elongated particles increase need for water and cement
Handling and Placement.	Aggregate can be separated from the cement; overvibration can lower the air entrainment or burst forms; undervibration results in poor concrete distribution; lost air means lower resistance to freezing and thawing; long waits in trucks cause concrete to pre-set; rapid loss of water causes shrinkage, loss of strength; freezing causes destruction of cement and eventual disintegration; top scum or laitance results in poor bonding of concrete with other substances; movement of forms causes cracking
Curing.	Premature removal of forms causes weak cement-reinforcement bond and poor strength
Waterproofing.	Water can corrode reinforcement causing loss of cement-steel bond

Aggregate used in concrete for earth sheltered houses should be clean and well graded and should not absorb water. Specifying only natural aggregate will assure its purity from lesser quality materials (such as shale, slag, slate, or clays). The amount of cement to be used in the concrete should be specified according to desired strength (for example, 4,000 pounds per square inch). Common ratios for earth shelters are five or six sacks of cement for each cubic yard of concrete. A direct relationship exists between the strength of the concrete and the amount of cement, and specifications should be met without exception.

The water used to mix concrete should be of drinkable quality and free of soil, oil, and other impurities. A very important factor in concrete strength is the ratio of water to cement. Adding excessive amounts of water to make the concrete easier to work is a common (and serious) error. Too much water weakens the cement

and reduces the waterproofing characteristics. The minimum water-to-cement ratio is about 35 to 40 percent or approximately 4 to 4½ gallons of water per sack of cement. It is not uncommon to add special chemicals called plasticizers to concrete in order to improve the workability without having to add excessive water. Plasticizers are expensive, however, and require precise control at the time of mixing to produce concrete that works and sets properly with the correct strength. Cylinders of the concrete are tested to determine concrete strength (see fig. 5-8), which should be around 4,000 pounds per square inch for earth shelters.

Concrete has to be thoroughly mixed so that aggregate is evenly distributed throughout the mass. Aggregate tends to settle to the bottom and the cement to rise to the top, weakening the

5-8. Concrete sample for compression strength testing. (Courtesy of Lon B. Simmons)

concrete. Concrete should be poured into forms with as little "bouncing action" as possible; the concrete coming from the chute should not ricochet off the form walls. Separation of aggregate from cement will occur unless the velocity of all components remains equal during the pouring and placing process.

After pouring, concrete needs to be further agitated (vibrated) to facilitate flow to all corners and to surround the network of reinforcement, electrical conduit, plumbing, and so forth. Care in the use of vibrators is essential since forms can be exploded from the force of the expanding concrete around a vibrator.

The arrival of concrete trucks should be timed so that the concrete does not undergo initial set before it is poured. Generally, concrete should be poured within 45 minutes from the time it is mixed. Waiting longer will require remixing and adding water with resultant weakening of the concrete. In this regard, a cement plant with radio-equipped trucks, located near the building site, is definitely preferred.

Well before pouring, forms are prepared to contain and shape the concrete. All forms must be exceedingly well braced and absolutely level or perpendicular. Wet concrete exerts signifi-

cant pressure on them, and if they are not properly braced, they will shift and bow, possibly, cracking the concrete. Ceilings, in particular, present difficulties. Moving or sagging of the shoring will produce cracked and uneven roof slabs. Secure, professionally engineered shoring is worth the expense of rental from construction-equipment rental companies.

Unreinforced concrete is strong in compression (as when a concrete bearing wall supports a load) but weak in tension as when a slab spans an open space such as a roof). For this reason concrete roofs, walls, floors, and other structural elements require steel reinforcement. The steel binds with the concrete to bear most of the tension load. Reinforcement bars should have rough surfaces that facilitate a strong bond between bar and concrete. The reinforced concrete must be supported at proper intervals or it will bend under its weight and that of other loads.

An engineer carefully considers the stresses placed on the reinforced concrete and recommends type, size, amount, and spacing of reinforcement. The quality of concrete (mix and water relationships) is also closely monitored. Reinforcement will be placed along the lines of greatest tension. For example, significant amounts of reinforcement bars (rebar) are positioned close to the bottom of a roof slab, around columns, and near the tops of beams. To limit cracking further, reinforcement is also used in areas subject to the stress associated with temperature and shrinkage. Figure 5-9 shows the reinforcement associated with the post-tensioned roof. In this type of structural system, stressed cables (stretched after the concrete has set) provide most of the reinforcement and tension strength of the concrete.

Structural strength is a prime concern in earth sheltering, and the problems can be many. Stresses near columns and supporting walls, floor and wall junctions, skylight openings, windows and doors, and drains all demand professional attention. In some cases, free-span roofs are required, while in others many support walls are appropriate. Cold joints (construction joints where concrete is poured

5-9. Reinforcement associated with a post-tensioned roof.
(Courtesy of Lon B. Simmons)

against cured concrete) need special handling. Obviously these technical points demand a professional's involvement in this most important aspect of the construction of an earth shelter.

Figures 5-9 and 5-15 illustrate post-tensioning, a reinforcing method for cast-in-place concrete earth sheltered homes. Benefits of post-tensioning over conventional cast-in-place concrete reinforcement and precast panels include increased strength, freedom from excessive cracking, reduction in material used, and more watertightness. In post-tensioning, plastic-coated and lubricated steel tendons are cast into the concrete. Because there is no bond between the lubricated steel in the plastic sleeves and the concrete, the anchored steel tendons can be pulled and stretched after the concrete has set. This squeezing action puts the concrete

in compression, yielding structures of impressive strength. The entire building is reinforced like a precast reinforced concrete plank.

The tendons are easily positioned in the forms to fit a variety of shapes (such as curves and domes). Tendons are usually run in a grid pattern through roofs, down walls, and across floors so that a very rigid, well-knit structure results. This strength allows clear spans of 28 feet or more.

Post-tensioning requires skilled personnel for controlling the quality of concrete as well as for laying and applying tension to the cables. Only good quality concrete can withstand the stress placed on the concrete by the post-tensioning process. Figure 5-10 illustrates the jacks used to pull the cable and shows a pulled cable with wedges in place. The ends of the cables are grouted to prevent corrosion.

5-10. Jack used to pull post-tensioning cables and a pulled cable with anchoring wedges. (Courtesy of Lon B. Simmons)

Once the cement is poured, it must have water and time to cure and harden. For at least a week after pouring, surfaces must be kept moist, out of direct sunlight, and protected from winds to prevent rapid evaporation. This can be accomplished on slabs by using commercial curing compounds, burlap bags, moist straw, dampened sand, or constant misting with a watering nozzle. For walls, forms should be left in place about a week. If the forms are stripped before this, the walls should be kept moist to permit curing and attainment of maximum strength.

Early removal of forms and premature loading will force the reinforcing steel to work before minimum concrete bond has developed around the reinforcement. Loss of bond sets the stage for early slab deflection in excess of allowances and could be cause for a building inspector to declare the slab dangerous. It is recommended that slabs be shored for around fifteen days to allow the concrete to develop the extra strength to bond firmly to the reinforce-

ment. Backfilling should not begin until after twenty-eight days to allow the concrete to reach its full design strength. During this time, the foundation tile may need to be partially backfilled to prevent silt from clogging it.

Foggy, damp days with temperatures in the sixties are best for pouring concrete. Hot winds tend to dry the concrete too quickly, and freezing endangers the ultimate strength of the concrete. For this reason, the weather should be closely scrutinized before deciding the final pouring schedule.

Although crack-free concrete is possible, even the best pours may develop shrinkage cracks caused by such uncontrollable factors as a slight shifting of forms or natural shrinkage of the concrete. Small shrinkage cracks usually do not go all the way through a slab or wall nor do they cause concern structurally. Cracks that permeate the entire slab are another matter; however, they can usually be avoided with close attention to concrete mixing, handling, curing, shoring, and formwork.

Preparing and Pouring the Floor

During form setting and slab preparation, the use of a surveying level (transit) will insure that all forms and surfaces are level (see fig. 5-11). Gravity drainage is desirable to allow any water around the slab to run off.

Many structural designs require that the concrete floor slab be free floating with respect to the concrete shell. This allows the foundations to move without buckling the floor slab. In other situations, the floor slab and foundations are poured at once and are connected (as in post-tensioning). The advice of the structural engineer is to be sought in making this decision.

When the forms are erected and securely braced for the pouring and screeding (leveling) processes, a good quality sand fill is poured as the bed for the slab. It should be screeded level

and power compacted (fig. 5-12). All footings must conform to the engineer's specifications, with all reinforcement properly placed. The slab should be protected from frost heave by a frost wall of extruded polystyrene insulation on the exposed side of the house.

Once forms are constructed and the sand bed is leveled and the footings prepared, the below-slab plumbing, electrical conduits, and similar items are ready to be installed. Subslab drain tile is also placed at this time. All of these items should be double-checked before the concrete is poured to be sure they meet specifications. All below-slab plumbing runs should be without seams (joints), and plastic sleeves be fitted around all water pipes that penetrate the slab. This prevents pipe damage with movements of the slab. Insulating hot water runs will probably prevent the cool earth from leaching

5-11. Use of transit to level forms of floor slab.

5-12. Use of mechanical compactor on fill sand for floor slab.

heat from the pipes. (If possible, keep all hot water runs as short as possible by placing use points such as baths and kitchens near water heaters.) If the drain pipes are to be used as earth tubes (see chap. 4), provisions must be made for connecting the drain pipes to the ventilation system of the home.

In general, plumbing should be kept out of all concrete except for the slab. If the plumbing and electrical system are run above the slab, routes and passageways should be planned in the walls and ducts. This will vary with the structural system used. Prestressed panels have hollow cores that are often used for plumbing and electrical runs.

Copper tubing is preferred for plumbing because of its strength, although plastic may be appropriate in some applications. Pressure-test all plumbing runs for leaks before they are covered with concrete. Water pressure should always be under 40 pounds per square inch.

The recommended slope for sewer lines below the slab is about ¼ inch per running foot. Too much slope means that liquids will move too quickly and leave solids behind. By carefully bedding sewer pipes, the effects of settling and soil movement will be minimized, especially where expandable clays are involved.

A vapor barrier of polyethylene plastic placed on the sand bed will prevent moisture from permeating up through the slab. This vapor barrier, plus any drain tile, should keep the area fairly dry. Insulation is placed below floor slabs only in northern climatic zones.

On the day of the concrete pour, enough people should be on hand to help distribute, vibrate, and work the concrete. The concrete should be of the right mix and should be kept at the right stiffness, or slump (the level of workability determined by measuring the amount of sag in a sample of concrete placed in a special cone (see fig. 5-13)). The engineer (or other professional) should be present to run slump tests according to accepted procedures. Plasticizers should be added only as instructed by the professional in charge of concrete specifications.

Table 5-2 highlights mistakes that can be made in the preparation and pouring of the floor slab, and figure 5-14 shows a floor slab being poured. Concrete should be thoroughly vibrated (especially around vent pipes, drains, and reinforcement). Wall-floor junctions are major areas of possible leakage and should be fitted with special strips of rubber that are designed to halt the migration of water. Called wa-

5-14. Pouring and finishing floor slab for earth shelter.

5-13. Slump test—the concrete should not collapse more than a specified amount. (Courtesy of Lon B. Simmons)

terstops, these rubber flanges are placed in the wet concrete of the floor where the walls will be poured. Entranceways and garage floors should be sloped to provide drainage.

TABLE 5-2

Common Mistakes Encountered
in the Pouring of a Floor Slab.

—fill material or soil base not compacted

—footings and column bases not properly sized or positioned

—reinforcement not properly positioned, wire mesh too close to the ground and not sufficiently embedded in concrete

—below-slab plumbing, drains, and electrical systems not correctly placed and marked

—vapor barrier not installed

—concrete not properly mixed, placed, or cured

—slab not level or not sloped for garage floors or entrances

Preparing and Pouring Walls and Roof

Sometimes it is possible to perform a monolithic pour, meaning that walls and roof are formed-up together (also the parapet or retaining wall) and poured at the same time. Monolithic pours add strength and have fewer seams but are also more difficult to perform than isolated wall and roof pours. Not all contractors have this ability, and only those who have experience with monolithic pours should be hired to attempt them, for mistakes in forming, pouring, working, and curing are difficult to correct.

Usually, however, circumstances dictate that walls and roof will have to be poured separately. Waterstops installed at wall-roof, roof-parapet, and other junctions are essential to forestall problems at these prime leakage areas. Furthermore, all surfaces at these junctures should be clean and properly reinforced to bind roof, walls, and floor together. Figure 5-15 shows the post-tensioning cables that tie all aspects of the house together. For homes using other structural systems, such as prestressed concrete panels on poured concrete walls, close attention should be paid to wall-floor and wall-ceiling junctions at the time of forming and concrete placement to ensure structural integrity and watertightness.

All electrical conduits and the forming for windows, doors, and other openings should be checked to make sure they are secure and installed with concrete-tight fittings. Particular attention should be paid to the location of switches, wall sockets, light fixtures, and utility fixtures (electrical lines, water pipes, and telephone lines) since they cannot be moved after the concrete has set.

Usually and most efficiently, a concrete pump (a truck equipped with a concrete delivery pump with a boom and long hose) is used to pump the concrete through the hose right into the forms for the walls and roof. A crane with a concrete bucket can also be used (see fig. 5-16). In pouring, make certain that the aggregate and cement do not separate by ricocheting off form walls and splashing excessively, or in any other way. Extreme care must also be exercised in the operation of the mechanical vibrator. Because the concrete expands around the vibrator, leaving this machine in wall forms for too long will cause the forms to bow out and split. Figure 5-17 illustrates the pouring and vibrating process on wall and roof.

5-15. Post-tensioning cables extending from walls and roof.

5-16. Crane and concrete pump used to deliver concrete to forms.

5-17. Pouring and vibrating of the walls and roof. (Courtesy of Lon B. Simmons)

Because in many cases the finished interior surfaces of the walls and ceilings will simply be the cured and painted concrete, the forms should be smooth. A careful search for contractors or suppliers of good quality forms is recommended since it is possible to achieve almost any desired concrete surface by the correct selection of forms. Figure 5-18 shows an artificial bricklike surface on a wing wall produced by a form surface.

As mentioned earlier, the shoring up and pouring of the roof is a particularly critical step in the construction process. The contractor and structural engineer need to work closely here. The shoring system must be professionally engineered and properly installed, with the correct amount of camber (slight upward crown) built into the system so that the roof levels out after the soil is placed on it (see fig. 5-19). Sup-

port walls and columns all need to be exactly positioned, and their reinforcement must be tied into the roof.

The roof deck must be formed and troweled to give the roof a slight slope, without depressions. Unless the pour is monolithic, the parapet and wing walls (if present) are poured later and carefully tied into the roof reinforcement. Waterstops are placed at junctions likely to leak.

It is recommended that wing walls be self-supporting rather than cantilevered off the concrete shell of the house. Cantilevering may cause cracks in the shell at the point of juncture if the wing walls move. Both wing walls and parapets should have good drainage behind the walls to prevent the buildup of excessive lateral stresses behind the walls.

For through-the-roof vents, steel pipes are

5-18. Brick-like surface produced by a mold that fits into the forms.

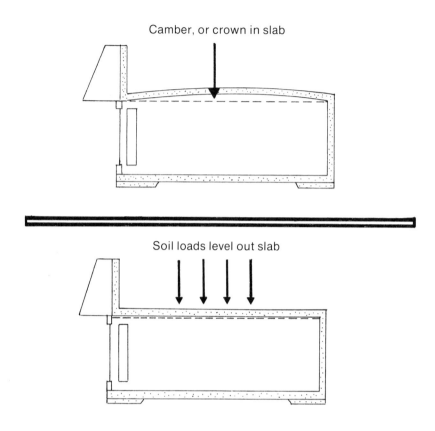

5-19. Roof-slab camber that levels after backfilling.

recommended over plastic since concrete binds to steel where plastic does not, and water leaks may occur around poorly bonded areas. Some designers advise no perforations through the roof, advocating that all vents be ducted out through walls to insure watertight roofs. Some waterproofing materials require certain kinds of surfaces for adequate adherance and this should be understood before the roof surface is finished off.

Other Structural Systems

Table 5-3 summarizes the concerns associated with precast concrete, sprayed concrete, concrete block, and wood construction. With precast panels, the building must be engineered and measured precisely to receive panels of exactly the right dimensions. Walls, reinforcement, bedding mortar, and other items in the building must be prepared and ready before the panels arrive. A suitably sized crane is needed for lifting panels off the truck, and enough trained manpower must be available for positioning the panels.

When cement is sprayed onto reinforcement (as with some domes and barrel shells), all the space around the steel rods and wire mesh must be completely filled. The pressure applications (sprayed concrete) perform well when done by a trained applicator with access to the right kind of equipment and materials. For domes and barrel shells, personnel experienced in constructing and loading such curved

TABLE 5-3
Concerns with Structural Systems

Structural System	Concerns
Precast Concrete	Opening of joints between elements
	Sealing of joints, cores, wall junctions
	Uneven deflection of elements
	Long lead time and high transportation costs
	Proper handling of panels
	Difficulty in cutting holes for skylights
Pneumatically Sprayed Concrete	Requires specialized equipment
	Possibly higher engineering fees
	Need to fill all spaces
	Application of insulation to curved surfaces
	Some waste of concrete
Cement Block	Easily cracked, allowing leaks
	Requires very good waterproofing
	Adequacy of structural strength
Wood	Need for chemical preservation
	Need for well-drained earth
	Not fireproof
	Adequacy of structural strength

structures should be employed. Applying exterior insulation to them can also be a problem. Usually insulation is positioned in the soil above the house.

The primary considerations with concrete-block construction are strength and watertightness. As the ultimate strength in a block wall comes from the steel reinforcement and concrete in the voids of the blocks, this must be done correctly to achieve adequate strength. Unfilled block walls will generally not be strong enough for earth sheltered applications.

The lack of thermal mass, susceptibility to fire, lower strength, and doubts about preservation are the main concerns with total wood homes. However, they do cost less and are easier to build than other types of earth shelters, and they probably experience fewer humidity problems. Waterproofing may be more problematic and care must be taken to use the correct nailing and reinforcement schemes.

Construction of Earth-Bermed Houses

Earth berms are apt to pose insulating problems because of the transition from above-to-below grade. When the below-grade walls are of concrete and the above-grade of wood, it is difficult to insulate uniformly and protect the area of transition. Rodents and other pests, as well as normal weathering elements, will attack this region more severely than in conventional homes. Lacking a reinforced concrete roof to help support the walls, earth-bermed homes also require stronger wall structure. Air infiltration and consequent heat loss is more difficult to stop in earth berms.

Waterproofing

The design aspects of waterproofing and insulation were covered in chapter 4. Here, a more practical discussion of techniques and materials will be discussed. To make an earth shelter watertight and comfortable, good surface and subsurface drainage, proper application of waterproofing materials and adequate, effectively placed insulation are essential.

Waterproofing is a detailed and specialized field that requires skills and equipment not generally available to the average do-it-yourself builder. Research and professional advice are therefore almost mandatory.

Figure 4-13 illustrates a generalized approach to the use of waterproofing materials. The waterproofing is placed directly on the roof deck, followed by insulation and a thin layer of gravel that is integrated with a system of drain tiles. The insulation protects the waterproofing and the structure from thermal shock and deflects some water, physical abuse, and root growth. The function of the gravel-covered drain tile and gravel layer is to relieve water pressure in the soil (see chap. 4). It is an excellent idea to provide a means for cleaning out the foundation drain tile as these tend to silt up after fifteen or twenty years. A strategically placed manhole or the use of tile with curved corners that allow for a rotary cleaning tool would be sufficient. There are drainage fabrics that can be used as alternatives to gravel, but they are generally more expensive. Most contractors and their supply outlets can furnish information on these fabrics and drainage mats.

Table 5-4 lists waterproofing materials and

briefly describes the areas of concern for each. The waterproofing layer is part of a total waterproofing system that includes surface and subsurface drainage, the waterproofing material, and the concrete or structural material.

TABLE 5-4

Concerns with Waterproofing Materials

Material	Concern
Asphalt and Coal Tar Pitch	Variability in quality of product Emulsification and lack of durability Brittleness and inability to reseal or span cracks underground
Vulcanized and Plastic Sheets	Surface adhesion and sealing of seams Water migration under membranes
Rubberized Asphalt	Surface adhesion Wrinkling after application Variability in quality of product High degree of skill required Water migration under sheets
Polyurethanes	Brittleness Loss of adhesion underground Formation of air bubbles Toxicity of fumes (isocyanate) Deterioration of polystyrene insulation by xylol fumes
Silicones	Shrinkage, cracking Evaporation losses of volatile substances Durability of silicone-concrete bond
Roll Goods	Stretching on vertical walls Adhesion to surface Water migration beneath membrane Sealing of laps and joints
Acrylic Latex	Emulsification underground
Cement Penetrants, Coatings, and Epoxies	No resealing qualities or elasticity Surface adhesion
Bentonite Clays	Swelling action can be hampered by salt or water impurities Need for good external pressure, soil compaction Inability to seal quickly after drying out Ineffectiveness where subject to running water

---

As discussed earlier, the concrete used in a structure should be a low-water, cement-rich mix, carefully placed to avoid surface powdering (spalling), cavities in walls (honeycombing), and irregular surfaces. The structure itself should be of tight, sound construction without excessive deflection and shifting. All chimneys, vents, and other penetrations should have good flashing with the waterproofing extending above grade. A slight slope to the roof will prevent ponding and direct water to the drain tiles.

Once the structure is completed and stable, waterproofing can be applied. To select the appropriate material, long-term objectives need to be considered, for some products cannot withstand the continual moisture, shifting, and temperature changes associated with long-term underground installation.

Theoretically, the best waterproofing material has a long life, is thoroughly warrantied, is easily applied, has a high leak-localizing capability, can reseal itself, is resistant to chemicals, yet compatible with such other materials as insulation.

Manufacturers usually warrant their products for one to five years, although most states require warranties in construction to be effective up to fifteen years. Rarely will this warranty include the extra costs of excavation or damage to exterior or interior materials. It should be remembered, however, that a guarantee is only as good as the financial state of the applicator or manufacturer, or their insurance coverage.

Easily applied materials offer better quality control since they are not as susceptible to error by construction personnel. Such a product should not be adversely affected by temperature extremes, concrete curing times, wind, or moisture, and should be able to cover vertical and curved as well as horizontal surfaces.

Since some leaks are likely in most earth shelters, water should not be able to travel beneath waterproofing. If a leak does occur, it should be possible to locate and repair it easily (preferably from the inside).

Awareness of the effects of chemicals in the soil on the performance of waterproofing materials is important. Some, such as bentonite, may be affected by salty soils. Also, the waterproofing materials themselves may adversely affect insulation, polyethylene sheets, caulkings, and sealants. Furthermore, some materials are used in combination with other waterproofing systems (such as bentonite and butyl rubber sheets), and the characteristics of each must be known in order to achieve the optimum results.

Dampproofing compounds such as concrete admixtures, polyethylene sheets, epoxy and acrylic paints, asphalts, pitch, and cement pargings are not impermeable to standing water and should not be used underground unless combined with other approved systems. If possible, it is wise to test the waterproofing by flooding before backfilling.

Of the materials listed in table 5-4, bentonite is most frequently specified by professional designers for earth sheltered applications. However, no product is able to cope with all situations, and professional advice is usually needed in the selection of appropriate materials and techniques.

BENTONITE

Bentonite, a natural volcanic clay, is presently a popular waterproofing material (see fig. 5-20). It is available as a powder (raw bentonite), in cardboard panels, or as a mixture with a variety of bonding agents added. Sprayed or troweled onto vertical or horizontal surfaces, bentonite waterseals a surface by absorbing many times its weight in water and swelling to create a pastelike impermeable gel. However, in salty soils, the clay does not absorb water well, and its performance is inhibited. When bentonite dries out, it pulls away from surfaces, cracks, and rehydrates slowly, which may allow for leakage until it swells again. Furthermore, running water, such as off a steeply sloped roof, may wash bentonite away.

Covering bentonite with polyethylene will help contain and protect it. Able to bridge small cracks and to reseal after being punctured, it also allows leaks to be localized and can be applied to nearly all surfaces as long as water is not present (which prematurely expands the clay).

LIQUIDS

Liquid-applied membranes such as the polyurethanes are either troweled, sprayed, or rolled on. The surface of the structure must be clean, smooth, dry, fully cured, and primed.

5-20. Bentonite in bag, applied on roof as powder, and troweled onto parapet wall. (Courtesy of Lon B. Simmons)

Trained applicators measure the thicknesses of the waterproofing layer because performance is hindered if layers are too thick or thin. Two coats are often recommended along with embedment of fibers in the first coat.

Potential problems associated with liquid-applied materials concern durability (tendency to become brittle over time), blistering and bubble formation, adhesion difficulties, and re-emulsification (latex compounds). Their use is not recommended on precast roof decks, masonry surfaces, wood decks, or other surfaces that have great potential for movement or cracking. Liquid-applied materials are most suitable for reinforced and post-tensioned, poured concrete slabs.

ROLL GOODS

Modified bitumens, sheet membranes, and rubberized asphalts are often referred to as roll goods. These materials contain asphalt and synthetic rubber combined with polyethylene sheeting. The rolls are usually 3 to 4 feet wide and are installed by lapping the sheets over one another (see fig. 5-21). Installation is difficult in low temperatures, and professionals are usually required to get good seals around roof and wall penetrations. A smooth, dry, and primed surface is prepared before application of most roll goods.

Butyl rubber and ethylene propylene diene monomer (EPDM) are prime examples of sheet membranes, which come in rolls up to 50 feet wide. These membranes are usually adhered to the surface or are loosely laid by professionals. Laps are sealed by solvents. The material itself is heavy, and because it stretches, its use on vertical surfaces is impractical. The major drawback of loose-laid sheet membranes is the difficulty in locating leaks because water can travel behind the membrane. Also, membranes warmed by exterior conditions may contract and pull away from edges when placed under cooler soil after backfilling; this contraction must be allowed for.

Insulation

The insulation of underground structures is still a relatively new endeavor and, at present, more an art than a science. The many differences found in soil (particle type, moisture patterns, densities, and so forth) make it difficult to predict how heat will flow in and out of a building covered with any particular type of soil. The principles that follow are based on state-of-the-art building practices, founded on presently held theories of heat transfer. Ongoing research in this area should extend our understanding, but for now, designers must necessarily rely on the reported performances of existing earth shelters.

Insulation, because of its low thermal conductivity, retards the loss of heat through the building envelope. Where concrete walls are covered with earth, externally insulated walls will retain heat longer and more effectively store heat gained from solar or thermal radiation. Concrete, while being a poor insulator, is a good heat storage medium if prevented from conducting its heat to cooler environments. The mean radiant temperature of a wall will remain higher for a longer period of time when the wall is externally insulated. If it is not insulated, the

5-21. Waterproofing with roll goods.

earth temperatures will more quickly influence the room temperature by withdrawing heat at a more rapid rate.

The situation changes if the earth sheltered building is made of low thermal mass materials such as wood. The insulation used here primarily retards heat loss but does not store it. Wood is also not as effective as concrete and masonry in promoting earth-contact cooling, for wood's thermal conductivity inhibits the flow of heat to the thermal sink of the soil.

In winter, heat is lost from a building by infiltration and conduction. Infiltration occurs when the warm inside air is replaced by the cold outside air. This loss can be controlled by sealing cracks and air leaks around windows, doors, and other openings. Earth sheltered houses are extremely tight structures, and the concern often is not enough air exchange rather than too much. Infiltration heat losses in an earth shelter can be controlled by purchasing high quality windows, doors, and external shutters and by giving heed to control devices for ventilation.

Through conduction, heat is transferred between two substances that are in physical contact with each other. Warmer molecules pass their vibrational energy to cooler, adjacent molecules. Conductive heat loss through windows can be reduced by adding curtains, blinds, and exterior shutters. Heat loss through walls can be controlled by placing insulation on the interior surface, in the wall itself, or on the outside of the wall. Table 5-5 and figure 5-22 give the advantages and disadvantages of these approaches to insulation in an earth shelter. The location of insulation depends primarily on the composition of the building's shell. If the building itself is to be used as a thermal mass to store heat received from solar radiation or auxiliary sources (for example, a wood-burning stove), then the insulation should be placed outside the shell. If the house is made of a light material such as wood, the main problem is heat retention and not storage, and insulation can be placed on the inside surface or in the wall. In this case, possible condensation of water vapor

in the insulation can be avoided by being sure to include a vapor barrier or venting holes.

TABLE 5.5
Advantages and Disadvantages of Insulation Placement.

Area of Installation	Advantages	Disadvantages
Within or inside structure	Cheaper insulation can be used. Safer from rodent and insect damage.	Thermal mass of wall not available to control temperature swings. Requirement for vapor barrier or venting because of possible condensation. Freezing temperatures may occur within wall.
Outside the structure but inside waterproofing layer	Insulation need not be able to withstand ground moisture conditions. Insulation will retain a higher insulation value because it is dry. Thermal mass of the building can be used.	Condensation may occur within the insulation if no vapor barrier is used on warm side of structure. Insulation must withstand ground pressure. Summertime cooling is reduced. Leaks may be difficult to prevent and locate. Thermal bleeds may occur at cold spots.
Outside structure and waterproofing	Waterproofing protected thermally and physically. Leaks less of a problem Thermal mass of building can be used. Extra insulation can be added.	Insulation must be moisture resistant and be able to bear earth load. Reduced summer cooling. Insects and rodents may attack insulation. Potential for thermal bleeds.
Within soil mass and detached from building	Increased summer cooling. Thermal mass of building and insulated soil can be used. Extra insulation can be added.	Placement of insulation may be difficult. Increased winter heat losses. Rodents and insects may attack insulation.

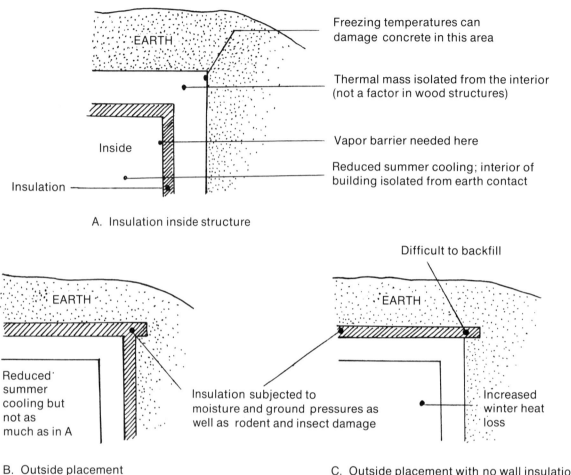

EARTH

Freezing temperatures can
damage concrete in this area

Thermal mass isolated from the interior
(not a factor in wood structures)

Inside

Vapor barrier needed here

Reduced summer cooling; interior of
building isolated from earth contact

Insulation

A. Insulation inside structure

Difficult to backfill

EARTH

EARTH

Reduced
summer
cooling but
not as
much as in A

Insulation subjected to
moisture and ground pressures as
well as rodent and insect damage

Increased
winter heat
loss

B. Outside placement

C. Outside placement with no wall insulation

5-22. Some disadvantages associated with inside and outside
insulation placement. (Source: U.S. Department of Energy, 1981)

There are as many variables in determining insulation strategy as in choosing the total building design. Heating and cooling demands will lead to different strategies. If cooling is the primary need, reduced insulation and heavy concrete walls may be recommended unless thermal mass is a problem (see chap. 4). If heat retention is the major concern, the options become more varied. Below-grade insulation, laid horizontally above the earth shelter (and enclosing a capsule of earth), appears to be more effective than an equivalent amount of the same material placed vertically next to the structure. Practical difficulties (such as how to place and backfill isolated insulation) must be solved,

however, and actual structures using this method still need to be evaluated over a period of years.

Table 5-6 gives recommended insulation R-values (resistance to heat flow) for different regions of the United States, with heating and cooling loads expressed in degree days (see chap. 3). The values given depend on climate, earth cover, and passive solar attributes, and the recommendations depend on the situation. The amount of insulation on the lower floor and wall areas varies according to the cooling load imposed by climatic and soil conditions. These portions of the building are of major importance in the promotion of earth-contact cooling.

TABLE 5-6
A General Guide to Recommended Insulation Values for the United States.

Region Example City	Degree Days (hdd/cdd)	Insulation Values (R-value)			
		Roof	Upper walls	Lower walls	Floor
Atlanta	3095/397	5–10	5	0–5	0
Boston	5621/127	20–25	10–15	10–15	0–5
Chicago	6127/241	20–25	10–15	10–15	0–5
Denver	6016/145	15–20	5–10	5–10	5
Houston	1434/998	5	5	0–5	0
Kansas City	5357/496	20–25	10–15	10–15	0–5
Miami	206/1045	5	5	0–5	0
Minneapolis	8159/160	20–30	15–20	10–15	5–10
New Orleans	465/772	5	5	0–5	0
Phoenix	1552/1554	5–10	5	0	0
Portland	5185/19	10–15	5–10	5–10	0
Salt Lake	5983/276	25–30	15–20	10–15	5–10
Seattle	5185/19	10–15	5–10	5–10	0–5

NOTE: In all cases, heating degree days are computed from a base of 68 degrees F and cooling degree days from a base of 78 degrees F; earth cover is 24 inches; walls are 10 feet high, and the floor area is not used as a passive solar gain storage area.

Figure 5-23 schematically illustrates the flow of heat out of a buried, uninsulated building in winter. The soil is settled around the structure, and the rate of thermal exchange is nearly constant. The solid lines represent the direction of heat flow. The closer the lines are together, the higher is the rate of heat flow. The highest heat flux is experienced through the roof and upper walls, and more insulation should be placed in those areas than beneath the floor and on the lower walls where less heat is lost. Areas buried deepest in the earth are near milder year-round temperatures. It is here that the balancing act between winter heat retention and summer cooling is best played out. Earth-contact cooling is promoted in the summer by the coupling of the earth with the building.

Table 5-6 should be used as a general guide for placing insulation. There are many situation-specific conditions that may require departures from this scheme. For instance it is often difficult to backfill over unsupported or "free-hanging" insulation panels, and it might be necessary to extend the insulation panels to the footing to gain support. Deeply buried domes also require different approaches.

In areas where freezing temperatures reach deep into the soil near the building, insulation to prevent freezing damage is important. Figure 5-24 shows where extra insulation may be needed and also reveals areas that can be avenues for exceptionally rapid heat loss. Concrete exposed to the elements can act as a heat leak or thermal bleed. Just as a small hole will drain a large vessel, thermal bleeds can draw off considerable amounts of heat. This can cause condensation damage inside by allowing the interior surface to fall below the dew-point temperature and to collect water.

Table 5-7 gives the dew-point temperatures for some commonly encountered situations.

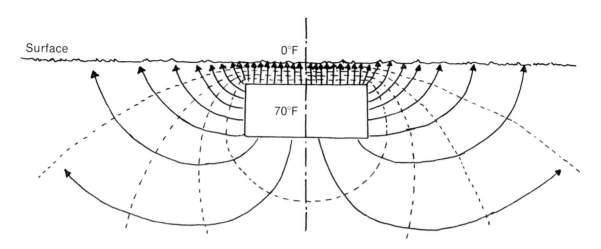

5-23. Heat flow out of a buried, uninsulated building in the winter at steady state conditions. Arrows indicate paths of heat flow through different temperature zones (dotted lines), the warmer areas being nearest the building. (Source: U.S. Department of Energy, 1981)

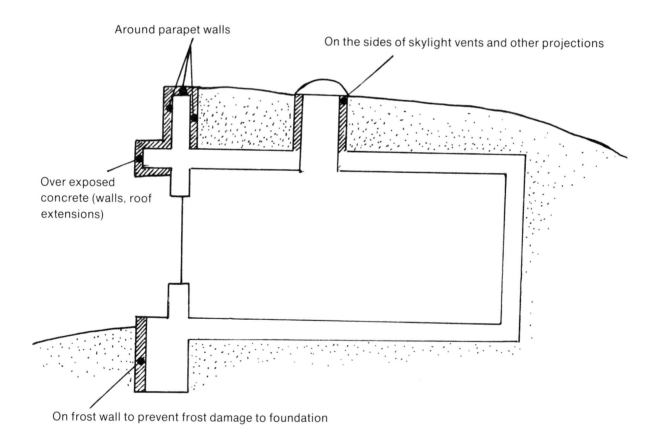

Around parapet walls

On the sides of skylight vents and other projections

Over exposed concrete (walls, roof extensions)

On frost wall to prevent frost damage to foundation

5-24. Frequently overlooked places where insulation is needed. (Source: U.S. Department of Energy, 1981)

Winter soil temperatures adjacent to the floor and lower walls of an earth shelter are normally above these dew-point temperatures. Tops of skylights, thermal bleed areas, and unprotected portions of the building may have problems and should be carefully insulated. Occasionally, insulation is stopped abruptly on a wall or roof, and a cold spot develops. Because these cold spots can cause more trouble than if the entire wall was uninsulated, avoid such interruptions in the insulation.

Condensation and subsequent mold and mildew development are most frequently encountered during the summer cooling season, which often has more relevance to earth sheltering than any other (see chaps. 3 and 4). Table 5-7 indicates that the potential for condensation is greatest when humidity levels are around 80 percent. However, if humidity remains about 50 percent, the wall surfaces in most earth shelters will not experience condensation. Humidity control techniques such as direct exhaust of humidified air, dilution through air movement, dehumidification, and the exclusion of humidity-producing activities are major lifestyle modifications.

TABLE 5-7

Common Dew-Point Temperatures for Earth Sheltered Walls

	Space Air Temperature	Room Relative Humidity	Dew-Point Temperature
Probable Winter Conditions	60°	30%–50%	20° F
	65°	30%–50%	20°–46° F
	70°	30%–50%	38°–50° F
	75°	30%–50%	42°–54° F
	80°	30%–50%	47°–59° F
Probable Summer Conditions	65°	40%–80%	38°–58° F
	70°	40%–80%	43°–63° F
	75°	40%–80%	47°–68° F
	*80°	*40%–80%	54°–73° F
	*85°	*40%–80%	58°–78° F

*High probability of condensation forming on wall.

Figure 5-25 reviews some of the insulating and passive solar heating principles mentioned in earlier chapters. If a floor, wall, or other component of the shell structure is to be used for passive solar gain and storage, external insulation will allow this mass to reach and maintain

5-25. The correct balance between earth cooling and heat retention strategies for the lower midwest region of the United States.

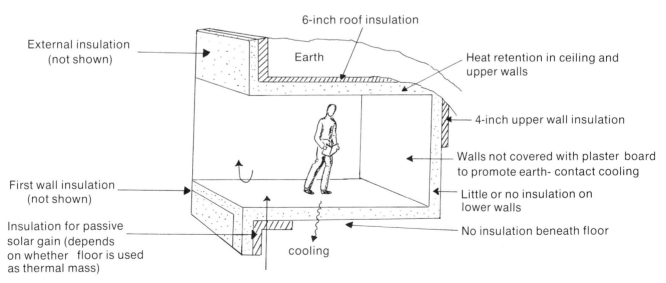

External insulation (not shown)

6-inch roof insulation

Earth

Heat retention in ceiling and upper walls

4-inch upper wall insulation

Walls not covered with plaster board to promote earth-contact cooling

Little or no insulation on lower walls

No insulation beneath floor

First wall insulation (not shown)

Insulation for passive solar gain (depends on whether floor is used as thermal mass)

cooling

Approximately 50 percent of floor is covered with tile or linoleum to promote cooling

higher temperatures. The interior finish of an earth shelter is important to the functioning of the thermal mass and earth-contact cooling. Concrete and similar materials will readily conduct heat away from warmer objects (such as hands and feet). If the concrete is covered with carpet, this conduction is significantly reduced, eliminating the "cold floor" feeling. Conversely, carpet and other coverings will inhibit cooling and thermal storage (see chap. 4).

As discussed earlier, the type and amount of vegetation growing on the roof and berms will affect the thermal performance of an earth shelter. This most intriguing component of an earth shelter, vegetation may be one of the most effective and economical aspects of the insulation plan.

INSULATION MATERIALS

Choosing the best insulation for earth sheltered applications is problematical because the insulation is usually placed outside the structure in contact with water, soil, and soil life. Figure 5-26 reveals some of the potential prob-

lems of exterior insulation. If the compression strength is unable to resist the weight of the soil and the force of soil movement, the insulation may flatten and lose the air pockets that give it a high R-value. In most instances, 20 to 30 pounds per square inch compressive strength is sufficient to resist most earth loads

Soil contains variable amounts of water and may at times be saturated for long periods. Water-resistant insulation, by not absorbing moisture, will retain its insulative capacity. If the R-value and thus the insulative properties are to continue to function, the insulation must also resist decomposition by soil solutions and waterproofing materials.

Table 5-8 lists the various insulation materials, describing characteristics that may be important to underground applications. It is not a definitive analysis, however. Extruded polystyrene (such as Styrofoam) generally outperforms expanded polystyrene (beadboard) and polyurethane in moisture resistance and the retention of its insulation capabilities (R-value). This is probably due to the difference in manufacturing.

5-26. Potential problems with insulation materials.

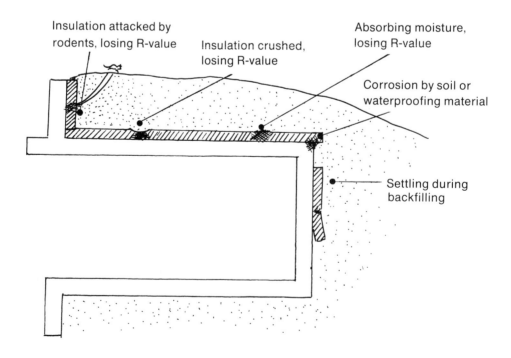

TABLE 5-8

Insulation Materials and Characteristics Pertinent to Underground Applications.

Material	Cost (1978) per bd. ft.	R-value per in.	Compression strength	Moisture resistance	Stability	Ease of Application
Extruded Polystyrene	$.16	5.0	good	Loses little of its R-value underground	Most resistant to damage by corrosion, insects, rodents	Tongue-and-groove configuration
Expanded Polystyrene (beadboard)	.11	4.0	adequate	Said to absorb moisture and lose approximately 30% of R-value	May be damaged by corrosion, insects, rodents	Commonly 2-by-4-feet panels
Polyurethane	.19	6.89	adequate	Said to absorb moisture and lose approximately 60% R-value	May be damaged by corrosion, insects, rodents	Can be sprayed to fit irregular contours

Extruded polystyrene is made by forcing a hot mixture of polystyrene solvent and pressurized gas through a slit. This extrusion and subsequent gas expansion results in a fine, closed-cell structure. Expanded polystyrene or beadboard is made by molding together polystyrene beads and then expanding them. The appearance of these two products is shown diagramatically in figure 5-27. There are over 175 companies making expanded polystyrene, and the quality varies. A quick test of quality is to break a piece of expanded polystyrene; good quality material breaks through the beads instead of around them.

Polyurethane, although having the highest initial R-value, is reported to lose a larger percentage of its insulation value as it ages and is exposed to moisture.

Which material should be used on earth sheltered structures? The answer is somewhat elusive. In its most recent design manual, the Underground Space Center recommends extruded polystyrene and advises against low density, expanded polystyrene for below-grade use when directly exposed to the soil. By using more expanded polystyrene, which is less expensive than extruded polystyrene, its insulation value may be increased. Protection and drainage should also be provided. Some under-

ground builders are using a combination: expanded polystyrene near the building and extruded polystyrene against the soil. Extruded polystyrene may also be the most resistant to rodent and insect damage.

The relative advantages and disadvantages of the various insulation materials may not be known until thorough, unbiased testing is performed and on-site results are published. Tests performed by the U. S. Army Cold Regions Research and Engineering Laboratory found that expanded polystyrene, buried under a highway, lost 30 percent of its insulative value after ten years, whereas extruded polystyrene lost only 8 percent, and polyurethane lost 57 percent.

For the present, extruded polystyrene appears to be the best choice for insulating earth sheltered structures. Expanded polystyrene should only be used where it will not be subject to long-term saturations. Polyurethane is the least suitable of the insulation materials for underground use. All forms of insulation must be installed so that they can drain. (Sometimes an interior drain will do.)

The use of a polyethylene vapor barrier over the insulation is also debatable. While the Underground Space Center recommends its use with expanded polystyrene, it has also been found that the plastic sheet will promote

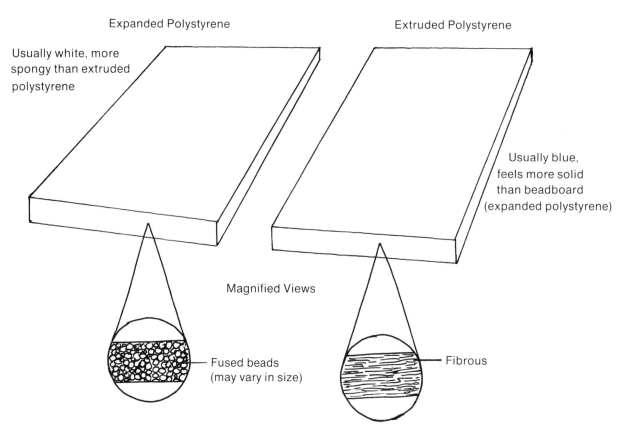

Expanded Polystyrene

Usually white, more
spongy than extruded
polystyrene

Extruded Polystyrene

Usually blue,
feels more solid
than beadboard
(expanded polystyrene)

Magnified Views

Fused beads
(may vary in size)

Fibrous

5-27. Appearance of expanded polystyrene and extruded polystyrene.

"vapor damming," which could decrease the R-
value by causing a buildup of moisture in the
insulation. Again, individual designers will
have to make this decision, based on knowledge
of soil conditions, water concentrations, and so
on.

When installing insulation several precautions are necessary. The insulation should not
be attached to the building in such a way that
the waterproofing will be damaged with settling
and movement, thus causing leaks at a later
time. As previously mentioned, effective insulation depends on proper drainage. When insulation is covered with soil, care should be taken
not to push and tear the insulation with the
backfilling equipment. As layers of soil are
added to the roof, the soil should be piled on
the insulation and graded back rather than

"bulldozed" or pushed, for this will tear the
insulation beneath the moving soil. (For a general insulating and waterproofing scheme, see
fig. 4-13.)

Insulation can be attached to the structure
with adhesives, nails, or backfill. The use of
nails or other attachments should not interfere
with the waterproofing, however. In backfilling,
the insulation (and polyethylene sheet) are held
against the structure with specially constructed
holding tees while soil is placed and compacted.

Backfilling and Landscaping

Table 5-9 lists the problems associated
with backfilling and some techniques to be considered. Improper backfilling will result in soil

settlement and damage to insulation, water-proofing, pipes, and utility hookups.

Drainage is disturbed when slopes and berms are not properly landscaped, while incorrectly backfilled topsoil adversely affects plant growth. Furthermore, careless use of heavy equipment can lead to structural damage to walls and roofs.

TABLE 5-9

Backfilling Concerns and Techniques

Concerns

—prevention of water and pressure buildup from soils and granular backfills

—prevention of settlement and excessive lateral pressures on walls and utilities by compacting fill

—maintenance of drainage slope

—protection of insulation and waterproofing

—prevention of structural damage from heavy equipment on roof and against walls

Techniques

—allow enough curing time for structure to gain maximum strength

—measure compaction and compact in layers

—avoid use of plastic clays as backfill

—avoid use of frozen clumps of soil or other debris as backfill

—place granular backfills strategically to lead water away from house

—avoid use of heavy equipment on roof and do not push soil on roof

—provide protection for insulation and waterproofing

Before backfilling, the structural integrity of the building should be established (see section on curing of concrete). Concrete should be cured, and maximum strength attained (usually twenty-eight days after pouring). All support structures (columns, buttresses, roofs, and so on) must be in place, and the waterproofing cured and ready for backfilling. Insulation panels should be in place unless they are to be secured by the backfilling process, in which case equipment and personnel should be ready and fully instructed.

Protective covering (heavy plastic, or cardboard) should be provided to shield the insula-

tion and waterproofing. Only small earth-moving equipment, such as the small loader in figure 5-4, are safe to use in placing soil around the walls and on the roof. Moreover, the soil should be evenly distributed in compacted layers around and on the building to prevent unequal loading (especially important in post-tensioned buildings). To prevent damage to the insulation and waterproofing, the soil cover on the roof should be graded back rather than pushed by the equipment.

Only appropriate backfill material should be used against the earth sheltered structure. Clunky soil such as frozen clay will settle unevenly, and expansive clays, especially if over-compacted, may cause structural damage to walls and roof. Sandy soils, due to the particle and moisture characteristics, make the best backfill, but gravel and specified soil types are also used. Backfill should be compacted in layers (8 to 12 inches thick) using appropriate equipment such as rollers or vibrating or impact compactors (see fig. 5-12). The moisture level and degree of compaction of the soil preferably should be measured by comparing it to standards such as the Proctor Test. The Proctor Test is a standard measurement of density of backfill used by engineers. The test is performed in a laboratory and involves compacting a sample of the soil to determine the degree of soil compaction and moisture required to support the building. Uncompacted material on slopes has been known to shift and slide creating depressions and cracks and damaging material below grade.

Walls to hold back earth near the front of the house are often needed (see fig. 5-28), and there are many ways to build them. Reinforced concrete requires a special design with cantilevered footings or tie-backs, excavation below frost level, and frequently, deep holes for drainage. Masonry (cement blocks mortared together) also requires footings and often develops cracks. Walls of rocks can be used as well as cribbing (railroad ties, concrete shapes, or commercially produced precast blocks specifically designed for use as retaining walls).

Reinforced earth is a patented retaining system using steel straps attached to earth shel-

5-28. Front and side view of a concrete retaining wall.

tered walls. The straps, laid in the backfilled earth in layers, reinforce the earth wall adjacent to the structure. The backfilled earth itself then becomes stable and is part and parcel of the wall. The backfill is free draining, which provides waterproofing. Possible corrosion of the straps is a concern, however, and this technique is now being used and evaluated in Europe.

General concepts and design issues in landscaping were discussed earlier (see chap. 4). Generally, the appearance and performance of any earth shelter is related to the selection of plants and landscaping techniques. Plants that are carefully selected for place and purpose can save energy, stabilize soil, beautify the home, and provide wildlife habitat. They are particu-

larly useful in solving one specific problem facing earth sheltered construction: how to stabilize slopes and berms. Wire or fabric mesh, concrete blocks, rubble or stones, along with plants, will serve this purpose. Figure 5-29 shows an earth berm stabilized with fabric mesh and plantings of honeysuckle.

Utilities

As discussed earlier in this chapter, wiring and plumbing routes in an earth sheltered house must be planned before the concrete is poured. This includes the location for any switch boxes and valves as well as utility points of entry into the house. These should be located

5-29. Slope stabilization using fabric mesh and honeysuckle plants on a steep berm.

in easily accessible areas. Accommodations for any future hookups, such as active solar systems or accessory water supplies, should also be planned as they are difficult to add later unless vents and other areas have extra space to accommodate pipes or other equipment.

An air intake tube is necessary for the wood-burning stove or fireplace. At least 6 inches in diameter, the tube is usually installed beneath the slab to provide combustion air in the winter and cool air in the summer as well as serve as an intake vent for general ventilation. A wall of considerable mass, such as a concrete bearing wall, will act as a heat storage medium if a wood-burning stove or fireplace is situated nearby. If the wood-burning stove will be used to preheat hot water, the water heater tank should be higher than the stove to facilitate thermosiphoning (see fig. 5-30). Water heaters should also have timers and other energy-saving features incorporated into the planning.

For ventilation and air movement, a system of fans and ducts should be installed to move air from room to room in summer and winter. In summer, a large, high volume fan (whole-house fan) is also needed to move cooler night air into the home as well as for general ventilation.

Because paneling on interior walls isolates the thermal mass, it is usually not practical to have extensive interior carpentry to hide electrical wiring and water pipes. Pipes and wiring can be led through holes in precast panels, cabinets and closets, through vents or ductwork, behind interior frame walls, in false beams, or along baseboards or other passageways. These routes should be preplanned, as concrete walls do not allow for many later alterations.

5-30. Wood-burning stove with water preheat coils and accompanying hookup to raised water heater.

Finishing the Interior and Exterior

Many designers recommend delaying the exterior work (fascia, siding, and trim) until the building has been backfilled and checked for leaks. Interior work should also wait until backfilling is complete and watertightness has been established.

Furthermore, some movement between floor and ceiling can be expected, and common sense dictates that a space (about one-half inch) be left at either the top or bottom of interior partitions. This will prevent the crushing of wooden partitions as the shell begins to distribute loads under the settling roof slab.

The interior of an earth sheltered house is finished much like a conventional home, although more attention is paid to potential sources of indoor pollution (see chap. 6). Concrete walls can be smoothed with a grinder (if smooth-surfaced forms were used this should entail minimal work) and painted using standard painting and interior finishing techniques (see fig. 5-31). All surfaces should be cleaned and primed with appropriate solutions specified by the manufacturer of the finishing material (stain, plaster, or paint). When uncovered, concrete walls act as thermal mass areas. Drop ceilings can be installed to accommodate wiring and plumbing, and all pipes should be well insulated.

Concrete stains and paints can be used to finish the exterior of an earth sheltered home as can various cementitious products, such as stucco. Alkali in concrete and mortar will degrade oils and oil-based materials, however, and only alkali-resistant coatings or latex paints should be used. It is best to consult a concrete

5-31. Finished interior concrete walls and ceilings in an earth shelter.

products supplier before using a product. As one recommended outside finish for the exposed front of an earth shelter, a layer of insulation is covered with a Fiberglas and stucco finish. The methods and materials involved in this procedure as well as companies that can be contacted are described in a *Solar Age* magazine article by E. Holland (see bibliography).

EXPERIENCES

Unfortunately, some of the best earth sheltered designs have been seriously compromised by the quality of construction. There is no substitute for experience in the handling of the details of excavation, concrete work, waterproofing, insulation, and other details. Many costly construction problems could have been avoided if the owners had hired only qualified personnel and followed appropriate techniques.

Excavation

Where the site was not subjected to a soil analysis, rocks, springs, and high water tables have been unexpectedly uncovered. Soils of low bearing capacity have also been discovered, demanding expensive countermeasures be instituted. In some areas soil has had to be trucked in from considerable distances because a high water table prevented digging on site.

The Minnesota Earth-Sheltered Housing Demonstration Program had problems with flooding, soil erosion, and hidden rocks. These caused considerable cost overruns (see Appendix D, sitework expenses).

Where owners hired inexperienced operators for excavation, many incurred considerable expense and loss of time in correcting errors. Other owners and architects reported considerable savings when the correct equipment was chosen and specific guidelines were given to operators for sparing trees, valued landmarks, and other site characteristics.

Deep excavations have led to excessively high soil pressures on walls and other structural members. This has also occurred with expansive clay soils (see chap. 3). Settling, un-

compacted soil has pulled away insulation and waterproofing, producing water leaks and cold spots. In one case, settling clay backfill caused a bituthene (rubberized asphalt) membrane to detach from a vertical wall, producing leakage and water damage. There are many cases where the careless use of heavy backfilling equipment has damaged insulation and waterproofing.

Granular backfills of sand or gravels, although recommended by many professionals, have occasionally produced cold walls in winter by allowing cold surface water to penetrate quickly to deeper areas adjacent to the earth shelter. Obviously, the use of granular backfills depends on climatic considerations as well as drainage. This area is currently being debated by earth shelter professionals. Surface water has also been directed up against homes by poor excavation.

Structural Problems

Although cave-ins are relatively rare with earth shelters (even in those not professionally engineered), some tragic cases have been reported. In one instance, a roof collapsed while being backfilled, injuring some of the workmen. In another case, the steel support columns buckled under the weight of added soil even though only one-third of the soil had been placed. Precast panels have come off edges of walls and collapsed when incorrectly placed or positioned. Cases of walls buckling and cracking and of roofs having excessive deflection because of poor design or mishandled materials have been reported.

In a two-story earth shelter that used a wood system for both the roof and the intermediate floor, no support was provided for the back wall, and it deflected after backfilling. The soil had to be removed and "dead man" anchorage (weighted cables) attached to the rear of the structure. This situation was caused by improper engineering analysis.

In the same study, another house was incorrectly constructed because the builder misunderstood the architectural drawings. Interior support columns for roof planks were placed too far apart, and the planks deflected excessively after placement. The planks were then removed, and an additional beam was installed.

There are other reports of excessive roof deflection attributed to the incorrect placement or omission of reinforcement materials. Although not all these accounts are in the published literature, their occurrence underscores the need for extreme care in the engineering and design phase.

Retaining walls have been a source of frequent problems in earth shelters. There are cases of inadequate structural design in these walls, which resulted in excessive deflection and even complete collapse under lateral soil pressures. In one instance, a retaining wall collapsed after a heavy rain. Another residence had to sacrifice some energy efficiency because the retaining wall could not support the specified amount of earth cover.

The concrete used in earth shelters has also been subject to damage. Many cracking and leakage problems have been traced to the addition of too much water or the separation of aggregate from cement paste. When water penetrates a concrete mass through cracks, the reinforcement will corrode, causing a poor bond to the cement and thus weakening the entire structure. Furthermore, the surface of improperly placed concrete will often flake or powder with poor handling, which prevents good adhesion of the waterproofing material. Plasticizers have been used to good advantage in keeping water content down.

Concrete poured in winter has frozen and lost its strength, while summer temperatures and winds have caused drying too quickly, leading to cracking and shrinkage in some cases. Where curing time is shortened, walls have cracked and moved under the pressure of backfilling. Inadequate shoring systems have shifted to produce cracks that penetrated the total concrete mass. Forms have broken under the weight of concrete or from overvibration, spilling concrete and creating much waste and added expense.

Little has been reported regarding pneumatically sprayed concrete (shotcrete), although in some cases, sprayed concrete has not permeated all areas of the reinforcement and honeycombing or sand pockets have resulted. In *Earth Shelter Living*, the High Pressure Shotcreting Corporation attributed much of the problem to whether the mix used for the shotcrete is wet or dry (see bibliography).

Total-wood homes have generally been well accepted, and satisfactory performance data has been gathered from them. Concern has been expressed over the potential toxicity of the wood preservative although preserved wood has been used for basement walls and other applications for many years. Most earth shelter professionals still strongly favor concrete, however.

Waterproofing

Possibly because water is a powerful solvent, able to bridge all but the most carefully designed and constructed defenses, it is not difficult to find problems with waterproofing materials in the literature (see table 5-10).

In most cases where dampproofing compounds (asphalt-based products) have been used as the primary waterproofing barrier, problems have been associated with the product's inability to bridge cracks associated with structural movement. Also, with time, asphalt has emulsified (gone into solution) with soil water. However, some earth sheltered homes have been waterproofed with asphalt for over twenty years without problems. The asphalts of twenty years ago were of better quality than those of today, for more of the long-chain molecules were left in the product. The same is true of coal tars, which today have toxic but valuable waterproofing molecules removed.

Besides being toxic to workers, polyurethanes do not bridge structural cracks well, nor do they adhere well to irregular or moist surfaces. In other cases, however, polyurethanes have worked very well. Roll goods (intact membranes) have also shown poor adhesion and have torn when used on vertical walls. In fact, many problems with waterproofing materials are caused by incorrect or inappropriate application. Either the concrete is too moist (some-

TABLE 5-10

Waterproofing Problems Reported in Literature

Waterproofing material	Reported problem
Dampproofing Compounds	Emulsification, brittleness Inability to seal roof penetrations
Polyurethanes	Bubbles in rolled-on applications Inability to span moving elements of wood and precast roof
Roll Goods	Lack of adhesion to dusty concrete roof Stretching and puncturing on vertical wall Slipping with soil settlement Being punctured by backfill materials
Bentonite	Leaks with cardboard panels Displacement with soil settlement Washing off steeply sloped roof Pre-expanding with water Drying out and not resealing quickly

times from condensation), or the surface is scaled, powdered, or irregular, or the material is not suited to the location.

Locating the source of a leak is often very difficult. Incomplete adhesion of waterproofing means that water can travel behind membranes. A leak in one location may actually originate some distance away. Because bentonite forms an impermeable gel that can reseal itself and bridge small cracks, thus allowing leaks to be localized, it is preferred by many designers. Even so, this product has had reported problems—again, mainly caused by incorrect application. According to many owners who have used it as waterproofing or to repair leaks, bentonite must be protected by a layer of polyethylene plastic to contain it and prevent drying out. In some cases, bentonite has pulled away from a surface because it has been inadequately contained, or it has washed away when subject to running water. One barrel-shell-type earth sheltered home reported difficulty locating and sealing leaks in a bentonite waterproofing system even though several excavation and repair attempts had been made. In dry climates, drought conditions have dried the bentonite, and it did not reseal quickly enough with the next rain to prevent leakage.

Recently, some bentonite products have had problems with asphalt or other binders.

These materials have either inhibited the swelling action of the clay particles or have separated from the bentonite mix and leaked through cracks into the interior of a building. This has led one manufacturer of bentonite products to declare Chapter Eleven bankruptcy when approximately 5 percent of its customers began experiencing leaks. None of the leaks is reported as being serious, and the company intends to reformulate the products and resume its services. Overall, however, bentonite appears to have considerable potential.

In many reports, the structure itself was the main cause of water problems because it was not built well enough to prevent the formation of large cracks or porous surfaces. In other reports, drainage was not provided (or the tile clogged or froze) and hydrostatic heads (water pressure) built up, exceeding the ability of the waterproofing to provide protection.

Insulation

Mistakes reported in insulating earth shelters range from installing no insulation at all to using the wrong kind. In the first case, the results of omitting insulation are cold walls that are very difficult to heat in winter and are subject to condensation in the summer. As mentioned in chapter 4, some earth sheltered homes take excessive amounts of wood to heat in winter but are fairly cool in summer (although some uninsulated homes report problems with mold and mildew). For northern homes, the omission of insulation under the floor slab has led to higher than suspected heat loss and has detracted from the performance of passive solar systems.

Thermal "nosebleeds" (areas of excessive heat loss by conduction through exposed concrete) are seldom noted in the literature, but many cases are described by earth shelter design professionals. Usually thermal breaks occur where wing walls meet structural walls or where concrete bearing walls extend to the exterior. They have also been reported around parapets (especially above earth sheltered garages), skylights and other penetrations (see fig. 5-26), and floor perimeters. Not many owners report serious concerns about such thermal leaks, but most have taken corrective measures such as applying external or (the less advantageous) internal insulation. Another source of heat loss, metal window frames, can be eliminated by replacement with wooden frames, which are less conductive to heat flow.

Certain types of insulation have become waterlogged, thus losing insulative value and impeding drainage, possibly leading to more problems with water leakage. Expanded polystyrene and urethane have both been found to absorb water underground. Consequently, most new construction is relying on extruded polystyrene (such as Styrofoam) for underground use. As stated in chapter 4, however, there is no general consensus about the correct type of underground, exterior insulation, and many designers are still using expanded polystyrene with few negative results.

Backfilling and Landscaping

Reckless backfilling has been the culprit in many insulation and waterproofing problems and has even led to structural difficulties. Many earth shelters suffer from the soil-settling, insulation-moving, and waterproofing-slipping syndrome. Reasonable care in using backfill equipment and compacting soil layers would have prevented most of the reported difficulties.

Many owners stress the importance of sloping the soil adequately on the roof to provide for good water run-off. Water should be directed away from the house, for some owners have reported heating problems when water is directed down against walls. Because of soil settlement and other alterations, many designers recommend delaying landscaping and final planting until soil-moving tasks are over. Most also report that native vegetation does better than introduced species, especially on dry, earth-covered roofs.

Utilities

As mentioned earlier, heating and cooling systems designed for above-grade homes have been found inappropriate for earth sheltered buildings. Such units are generally oversized for underground needs. The central thermostat system of conventional homes has not functioned well in the earth sheltered environment where a "zoned" system, with control systems based on regulating individual rooms rather than the house as a whole, does best. Also, the placement of heating registers needs considerable attention. A Minnesota study revealed that heating vents placed near south-facing windows lose considerable heat and detract from system efficiency. The same study also called into question the standard methods for placing forced-air ductwork. Earth shelters may require a complete rethinking of these systems.

Firsthand Experience

Not surprisingly, hindsight has revealed the aspects of construction that I feel are most critical in avoiding major mistakes. Regarding excavation, we were fortunate to have a near-perfect site and to find highly skilled operators. We uncovered some large limestone rocks, but these were easily handled with the front-end loader. Excavation was limited to a depth of 8 feet, with most of the digging less than 5 feet (see fig. 5-3). The soil was carefully monitored, and a post-tensioning structural system was chosen to accommodate the clay substrate. Thanks to the slope of the site, drainage problems with run-off water were negligible.

The post-tensioning required careful supervision by the engineers. The laying of cables was precisely planned, and specifications were closely followed regarding their positioning and reinforcement (see figs. 5-9, 5-10, and 5-15). The concrete for the floor and footings was poured simultaneously; the walls were poured using a crane; the roof required a concrete pump (see figs. 5-12 and 5-14). Concrete of 4000 pounds per square inch (we used six and one-half bag mix) was vibrated into place (fig. 5-15). The shoring system was engineered for the slab thickness specified. After the concrete had been allowed to cure, the cables were pulled using jacks (fig. 5-19), and their ends were grouted.

Waterproofing consisted of premolded membranes (fig. 5-21) and bentonite on the roof with cementitious and mastic coatings on the walls. Tie channels (wires in forms) on the walls were covered with a cement waterproofing, and waterstops were installed at the floor and wall junction. Drain tile was placed along the base of the footings, under the slab, and along the base of the parapet wall (see fig. 5-6). Expanded polystyrene, which was used as insulation, was covered with a layer of polyethylene plastic for protection and water deflection. (The expanded polystyrene is still in good shape and shows no sign of deterioration.) Over living areas, six inches of insulation were put on the roof, four inches on the upper walls, narrowing to one inch down to the footing. Two-inch-thick insulation panels were placed on the garage roof, with one inch of insulation on the walls. No insulation was placed beneath the floor slab.

Sandy-clay soil was used for backfill with 3 feet sloping to 2 feet on the roof and a long slope leading to a swale coming off the sides of the house. Topsoil was placed on the surface, and natural bluestem grasses were established, transplanting plugs of grass from the surrounding prairie.

RECOMMENDATIONS

Table 5-11 summarizes major mistakes that have occurred in the planning and construction of an earth sheltered home. Most can be avoided

---

TABLE 5-11
Avoidable Mistakes in Construction

1. No written agreements with designers and contractors that specifically designate where responsibility lies for leak repair, soil handling to replace settlement, crack repair, electrical repairs, and so on.

2. Failure to get good engineering and detailed adherence to specifications by contractors. This is particularly important for reinforcement choice and placement.

3. Poor concrete work (water-to-cement ratios, placement, and curing).

4. Improperly treated roof penetrations and floor-wall, wall-roof, and parapet-roof junctions and transitions.

5. Poor choice and installation of waterproofing material.

6. Poor choice and installation of insulation.

7. Inadequate air movement, and heating and cooling systems.

8. Careless backfilling and soil compaction.

---

by hiring competent personnel and providing for close supervision. Furthermore, contracts should define responsibility for any repairs or problems that might occur.

The following items demand special attention if mistakes in construction are to be avoided.

1. Slope and drainage need to be facilitated by appropriate excavation and backfilling.

2. All reinforcement and support elements should be professionally engineered and their construction closely supervised (and double-checked before concrete is poured). Moving and shifting structural elements are major causes of water problems.

3. Concrete must be of low water content and placed and vibrated correctly.

4. A complete waterproofing system (materials and drainage) should be professionally designed and installed with utmost care. Dampproofing materials should be avoided and strong consideration given to combining bentonite with other membrane systems such as EPDM sheets. Minimal roof penetrations, carefully waterproofed, are recommended. Plastic pipes should not be used for roof penetrations as concrete does not adhere well to plastic.

5. Choose extruded polystyrene over expanded polystyrene if economics permit. Expanded polystyrene may be acceptable in dryer climates but not in wet regions. If used, it should be protected with a layer of polyethylene plastic.

6. All potential thermal bleeds should be insulated (see fig. 5-24). This includes chimneys, which can cause considerable condensation problems unless insulated and protected.

7. During backfilling, only small equipment should be allowed on the roof, and all soil should be carefully placed and compacted with topsoil on the surface. All interior and exterior finish work should be postponed until the structure is deemed watertight.

CHAPTER 6

LIVING IN AN EARTH SHELTER

By wisdom a house is built, and by understanding it is established; and by knowledge the rooms are filled with all precious and pleasant riches.

Proverbs 24:3–4

Because an earth shelter is airtight and made of earth-associated construction materials, it may seem to be a "closed-in" environment. However, if adequate daylighting and window space are provided, this need not be the case. Also, earth sheltered houses may accumulate more interior air pollutants than conventional homes. For this and other reasons, the occupants of earth shelters and other energy-efficient homes must be more aware of the state of the exterior and interior environment than those who live in open, aboveground dwellings. To maximize performance of an earth shelter, standard energy conservation practices and timely operational activities must be observed. This chapter will discuss possible problems with the earth shelter environment and their solutions.

PRINCIPLES

While there is an initial bias against underground living, most people do feel comfortable in the earth sheltered environment. The public seems to be unduly influenced by the idea of a "damp, dark, cave," an attitude that is quickly dispelled by education and exposure to earth sheltered homes.

Indoor Air Quality

The same features that make a building energy-efficient—the use of thermal mass and the reduction of infiltration—can also lead to problems of indoor air quality. Recent studies have found relatively high levels of harmful pollutants in airtight and energy-efficient conventional houses, nor are earth shelters immune to such problems (see table 6-1).

These pollutants fall into four categories: dust and particulates; toxic gases; formaldehyde and other organics; and radioactive substances, such as radon and its decay products. Figure 6-1 illustrates how these substances can enter and accumulate in an earth shelter, and table 6-1 gives their sources and possible health effects.

Activities such as smoking, cleaning, and woodworking create dust consisting of small particles and fibers of smoke, silica, clay, biological matter, lint, and metal. Dust particles smaller than 2 micrometers in size are most likely to be inhaled and retained in the lungs.

Gas ranges and other appliances that produce heat through open flames are a source of such hazardous gases as carbon monoxide and nitrogen dioxide. Wood-burning stoves and

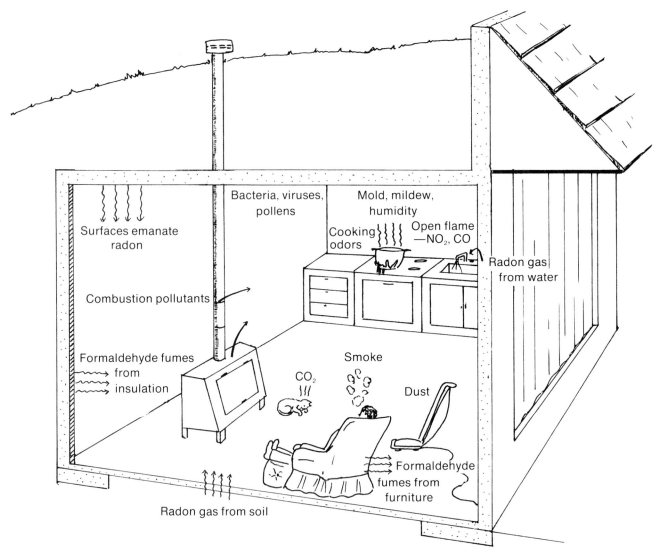

6-1. Potential indoor air pollution problems in an earth shelter.
(Source: U.S. Department of Energy, 1981)

fireplaces are also pollutant sources, generating carbon monoxide, nitrogen dioxide, sulfur dioxide, formaldehyde, acetaldehyde, phenols, benzopyrene, and dust particles.

Table 6-2 describes the odors of some of these compounds. Most pollutants, however, can be at dangerous concentrations without being noticeable. Decreased resistance to respiratory infection and other diseases mentioned in table 6-1 may be the first signs of their presence. For example, angina or chest pains may be the first sign of carbon monoxide poisoning.

The major sources of organic indoor pollutants are urea formaldehyde foam insulation, particle board, plywood, fabrics, tobacco smoke, and indoor combustion. The "new smell" of some furnishings comes partly from some of these substances. In babies, formaldehyde has caused vomiting, diarrhea, nose bleed, and rashes, while headache, tiredness, nausea, and diarrhea have been noticed in adults. Odors from cooking, wastes, smoking, and building materials can be simply annoying, as can problems associated with excessive humidity in

TABLE 6-1
Some Common Indoor Air Pollutants In An Earth Sheltered Home

Pollutant	Description	Typical Concentrations and Remedies
Radon	Radioactive gas from earth's crust; found in soil, building materials, well-water, and natural gas; may cause lung cancer.	Earth shelters (1.2 to 10.8 pCi/l) Conventional homes (0.9 to 17.6 pCi/l) Concentrations above 4 pCi/l are of concern. To reduce exposure increase ventilation or use heat exchangers, seal all cracks (also see table 6-6).
Formaldehyde	Strong-smelling gas found in some insulation and in various building materials and other products; causes nose, throat, and eye irritation—possibly nasal cancer.	60 to 1,673 parts per billion (463 parts per billion average); use low-formaldehyde materials such as particle board and plywood, increase ventilation.
Carbon Monoxide	Colorless, tasteless, odorless gas produced in all fuel burning; lung ailments, impaired vision, and brain functioning—fatal in high concentrations.	2.5 to 28 parts per million; vent all stoves, install exhaust fans above gas stoves and keep stoves properly adjusted, clean chimneys, do not let fires smolder, do not idle car in garage, increase ventilation.
Nitrogen Dioxide	Colorless, tasteless gas formed during combustion; causes lung disease after long exposure.	0.005 to 0.11 parts per million; remedy same as carbon monoxide.
Combustion Particles	Small smoke and organic particles from combustion processes; may cause lung cancer, emphysema, heart disease, respiratory infections.	10 to 70 micrograms/cubic meter (benzo-(a)-pyrene 7.1 to 21.0 nannograms/cubic meter); avoid smoking tobacco, seal leaks in woodstoves, vent all combustion appliances, change air filters regularly.

TABLE 6-2
Odors of Some Pollutants

Pollutant	Odor Description
Acetaldehyde	pungent and sweet
Acetone	sweet, pungent, fruity
Ammonia	pungent, like urine
Butyric acid	sour, rancid
Carbon tetrachloride	sweet, pungent, like petroleum
Formaldehyde	strawlike and pungent
Hydrogen sulfide	like rotten egg
Phenol	medicinal
Styrene	like rubber, sharp, sweet
Toluene	like mothballs, sour, burnt

earth shelters. Humidity levels build up from introducing humid outside air, cooking, bathing, and the like. Spores of molds and fungi multiply in humid conditions and can cause disease.

Radon has recently been the subject of intense interest among earth shelter professionals. This worldwide gaseous trace element, commonly found in soils and rock formations, is radioactive. As it breaks down, radon–222 is formed along with other decay products. These gases are colorless and odorless but are suspected of being significant contributors to the development of lung cancer, especially when in high concentrations such as might occur in airtight homes.

These radioactive substances enter the respiratory system on dust. Once in the lungs, they can cause body cells to become cancerous. Radon levels are measured in picoCuries per liter (pCi/l). One pCi yields 0.037 radon atoms per second. Common levels in homes are around 4 to 6 pCi/l. High levels (greater than 4 pCi) may cause more than 10 percent of the lung cancers in the United States although it is difficult to cite reliable estimates.

Most radon in dwellings seeps from the soil through cracks and openings in the structure of a home. Occasionally, these radioactive gases are sucked out of the ground by the home's ventilation system. Radon can also come from the stone, gravel, or sand used in construction. Because earth sheltered structures are made of crustal materials such as concrete, they may receive extra radon doses from these materials.

Tap water and natural gas can also carry radon into a structure.

Radon levels may fluctuate over time depending on the activities in a home. High levels may occur when the house is tightly closed or when radon is pumped from the soil by suction (such as switching on a vent fan in the bathroom or kitchen). Conversely, lower levels will be found when the house is open and freely ventilated.

By using commercially available services, radon levels can be monitored, although it is difficult to get reliable readings on all radioactive substances associated with radon. Table 6-1 gives some representative levels of radon in earth sheltered and conventional homes. (More experiences with radon and other household pollutants are discussed later in the chapter.)

Obviously, measures must be taken to control the concentration levels of pollutants in an earth sheltered home. Radon, formaldehyde, and other pollutants can be largely eliminated by preventing their entrance. Products that contain high levels of formaldehyde and exude a strong "new smell," such as particle board, fiber board, plywood and laminates, and foams for thermal insulation should be avoided. This also applies to the use of building materials high in radon concentrations. Basalts, sandstones, and many limestones have lower-than-average radon whereas granite tends to contain high radium levels. Great variability exists in the radium concentrations regionally, and it may be wise to check with regional experts associated with the U.S. Environmental Protection Agency for specific recommendations.

Cracks in the structural shell of an earth shelter should be sealed, and the waterproofing layers should be continuous and intact to prevent radon entry. Likewise, pipes and other penetrations should be carefully waterproofed and sealed. Plastic vapor barriers can also be installed below slabs.

Natural ventilation that brings in outside air can be very effective not only in regulating temperature and humidity but also in lowering the levels of indoor pollutants. Earth shelters, because of their thermal stability, may allow more air changes with less temperature drop than would be possible in homes of less thermal mass. Timers and automatic switches can be installed on equipment to assure ventilation at correct times, especially in closed-in areas such as closets. Mechanical filtering, dehumidification, and air cleaning will also help as will the use of properly installed heat exchangers.

Controlling the Thermal Environment

Because the indoor environment of an earth shelter is influenced by the cycles of sun, earth, and seasons, the occupants must be attuned to the status and availability of these natural energy sources if their home is to operate at maximum efficiency.

An earth shelter will not perform as efficiently its first year as it will in later years. It takes months for the soil and underground structure to reach a stable thermal equilibrium. For the first year after backfilling, an earth shelter may require much heating and cooling from the auxiliary systems because of the unequal distribution of temperatures between the house and recently disturbed soil. Furthermore, the temperature of the huge thermal mass of the building and surrounding soil has to be raised to comfortable levels. Good energy management can then keep this "thermal flywheel" in the correct temperature range. The growth of vegetation and soil settlement around the earth shelter will also have important long-range consequences.

The curing concrete of an earth shelter will lose significant amounts of water that will elevate humidity levels for the first year or two. An 8-inch concrete wall, for example, will lose about one pound of water per square foot of area, at 50 percent relative humidity. The time of backfilling also affects this relationship. Earth shelters covered during the late fall, winter, or early spring generally experience cooler temperatures and higher humidity than those backfilled during the warmer, drier months.

The earth shelter owner should expect to carry out different building management procedures during the first year or two than during the remainder of the building's life. More auxiliary heat and dehumidification may be necessary while the soil is reestablishing its thermal characteristics and the building is losing moisture.

The elevational earth shelter diagrammed in figure 6-2 will be used to explain the operational procedures necessary for maximum thermal performance. Although this type of earth shelter is somewhat unusual, its simple design

6-2. Three views illustrating the energy systems of a passive solar, earth sheltered residence. A. Earth covering. B. Trombe walls. C. Earth tube (air inlet for wood-burning stove). D. Windows for passive solar heat gain. E. External rolling shutters for sun control. F. Air-lock entry. (Designed by Lon B. Simmons)

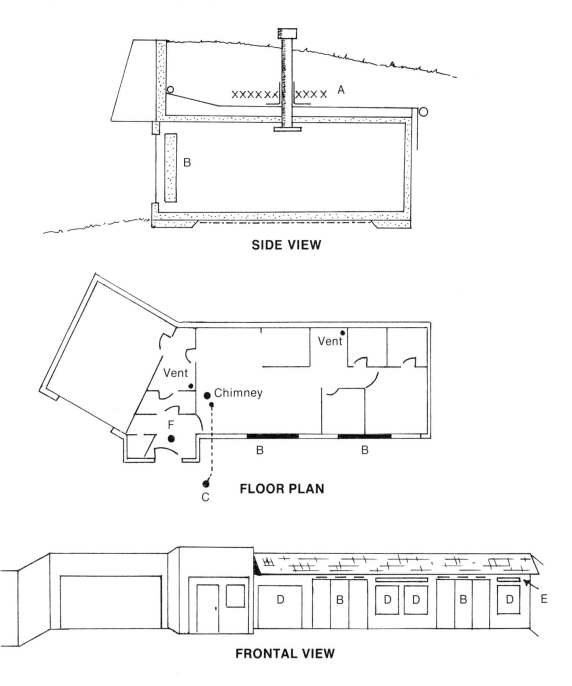

and multiple energy systems clearly illustrate the operational procedures that most owners of earth sheltered homes will have to perform once thermal equilibrium has been established. It is earth covered and equipped with: two Trombe walls with controllable vents; windows and patio doors for direct solar gain; exposed concrete for primary and secondary thermal mass (see chap. 4); exterior rolling shutters; a 10-inch-diameter, below-slab air intake tube with vent-controlling dampers; a whole-house exhaust fan; a ceiling fan and duct system; external venting in both bathrooms; an air-locked entryway; and wood-burning stove with hot water preheat coils. In addition to these systems, the home has a movable summer trellis for shading and high-crowned deciduous trees on the south side (see chap. 3).

Table 6-3 lists the seasonal operating procedures for the systems in the house. This operational regime is for a home located in the midwestern United States but should apply to homes located elsewhere, assuming that modifications based on local climate have been incorporated.

To operate the energy systems of an earth shelter effectively, the owner needs easy access to conveniently located thermometers and relative humidity meters, both inside and outside the house. Inexpensive thermometers and hygrometers are available at most hardware stores, while more expensive recording hygrothermographs can be purchased from scientific supply companies (see fig. 6-3). The latter allow a con-

6-3. Hygrothermograph for recording temperature and humidity fluctuations.

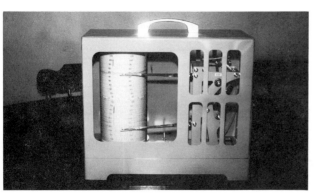

TABLE 6-3
Seasonal Operation of Energy Systems

Energy System	Operation
Trombe Walls	Winter: open inside vents during the day and close at night; raise exterior shutters when sun is available; lower at night or in cloudy weather; exterior summer vents should be closed.
	Summer: open inside vents and summer vents; lower shutters for duration of hot weather.
	Spring and fall: operation is flexible depending on needs; most of the time, the Trombes can be in the winter mode.
Windows	Winter: raise exterior shutters when sun is available and close at night or on cloudy days; open windows periodically to facilitate ventilation but keep closed most of the time.
	Summer: lower exterior shutters when no shade is available during the day and raise during night; open windows when outside temperatures and humidity levels permit, usually at night.
	Spring and fall: operation is flexible depending on needs; most of the time the windows can be open during the day and closed at night.
Air Intake Tube	Winter: open vent controls just enough to allow combustion air for stoves and air exchanges.
	Summer: open vent controls when incoming air temperatures and humidity levels are near comfort levels and close when such is not the case; usually vents are open only at night.
	Spring and fall: operation is flexible; usually vents are open.
Fans	Summer: use whole-house ventilating fan at night to "flush" the house with cool and dry air; use ceiling duct-fan system and smaller portable fans to provide air movement.
	Winter: do not use whole-house ventilating fan; seal off its port from the inside; use ceiling duct-fan system to distribute heated air.
	Spring and fall: use natural ventilation.
Roof Vents and Chimneys	Winter: use dampers to provide enough air for ventilation and to exhaust odors and pollutants.
	Summer: keep dampers fully open to facilitate air flow.
	Spring and fall: same as summer operation.

tinuous, permanent record of temperature and humidity fluctuations, thus giving added information that may be valuable in evaluating long-term performance and identifying trends and relationships between house and environment. Indoors, instruments should be located centrally and toward the back wall; outside, instruments should be placed in the shade, easily readable from inside the house.

During the winter, it is important to admit solar radiation and keep out cold air. This entails raising shutters to expose windows and Trombe walls at sunrise and closing vents and other openings, including the summer vents on the Trombe walls (which will be closed all winter) and the air intake tube. However, there must be enough fresh air for woodstove combustion, breathing comfort, and prevention of a buildup of indoor pollutants. At sunset, the exterior shutters should be closed, and the wood-burning stove or other auxiliary heating source can be used. Interior heat-producing activities, such as cooking, can be scheduled to supplement this heat input.

Extreme care should be exercised in the installation and operation of the wood-burning stove. Incorrectly installed, woodstoves have caused countless fires, and many deaths have resulted from poisonous gases (carbon monoxide) released from improperly operated and vented stoves. Small wood loads, added frequently to hot fires, are best for controlling pollution, lessening creosote buildup, and increasing efficiency.

In winter, the auxiliary heating units will reduce humidity levels in earth shelters to around 30 percent, which is near the recommended levels for winter conditions. For outdoor temperatures between 0° and 20° F, recommended indoor relative humidity levels are 25 to 35 percent. For temperatures of −20° to 0° F, relative humidity levels of 15 to 25 percent are recommended.

By spring, the cold winter temperatures will just be reaching the walls of the earth shelter. Because of this thermal lag, it may be necessary to use auxiliary heating slightly longer than in an aboveground home. However, if solar radiation is available, the incoming heat from the sun will more than compensate for thermal lag. Usually, spring is an easy time for most earth shelters, especially in the Midwest.

The cool walls will enhance comfort well into the summer when it becomes necessary to block out solar radiation and high outside humidity levels. Overhangs, shade trees, and outdoor shutters will prevent most of the sun's heat from entering the home. Temporary trellises or sun screens will increase the shading effect. Humidity can be controlled by the judicious opening and closing of intakes and vents.

When outdoor humidity levels permit, summer vents in the Trombe wall (see fig. 6-2) should be opened, along with the air intake, woodstove, and bathroom vents. Excessive intake of outside air must be closely watched, however, since temperature and humidity problems can result when hot and humid air is brought into an earth sheltered home. Because subjective human judgments often are not accurate, thermometers and relative humidity meters should be consulted before deciding to ventilate.

A major goal in summer, admitting cool air at night and retaining it during the day, is easily achieved by the earth shelter, with proper operation particularly of the whole-house ventilation fan at night. Cool air is retained during the day by closing windows and vents in the morning. In some climates, this is relatively easy although in others night conditions may be limiting (see chap. 3).

Objective readings from thermometers and humidity meters allow the most efficient use of fans, air conditioners or heat pumps. Most earth shelters require circulating interior air to promote convective cooling and air movement, and portable fans and air-handling systems do this quite well. Many earth shelters have temperature and humidity sensors that automatically switch on fans, air conditioners, and exhaust systems as needed. These controls must be checked periodically, for incorrect functioning results in a major energy waste.

In autumn great vigilance in the operation of an earth shelter is usually unnecessary because outside temperatures are more moderate and night conditions can be used effectively to provide comfortable daytime living.

Should an earth shelter be left unoccupied

for any length of time (such as during vacations), precautions against air stagnation should be taken. Usually this involves the periodic operation of dehumidifiers and air-handling systems as well as the introduction of fresh air. Closed-in areas, such as closets, should be left open to allow air movement. Also, it is a good idea to promote air movement under such items as mattresses, couches, dressers, and other places where mold, mildew, or other odors may accumulate.

Controlling Energy Use

Clearly, the occupants of an earth shelter can greatly influence the thermal performance and comfort of their home. The same is true for energy use. With the exception of fans, air conditioners, heaters, and dehumidifiers, energy requirements of an earth shelter are generally met by solar input and earth sheltering. Therefore, most of the auxiliary energy use of an earth shelter is related to household appliances. Table 6-4 lists the typical usage patterns for some of these appliances in an average American home. Usage usually peaks around mealtimes, with the highest overall peaks occurring in the early evening.

Because natural gas, propane, oil or other open-flame-type fuels may contribute to air pollution or safety problems, electricity is the most common form of energy used in underground homes. Table 6-5 lists some ways to reduce its use. Also, by solar heating of water or by preheating it with wood-burning stoves, electricity can be saved. These two systems can make valuable additions to earth sheltered designs (see chap. 4).

Following the guidelines in table 6-5 will have an impact on energy savings. Most experts agree that the living patterns of the residents are significant in determining the overall energy performance of an earth shelter. To demonstrate this, subject families have purposely been extravagant with energy use one day and conservative on another. When compared, the differences are impressive, suggesting that savings of hundreds or thousands of dollars can be achieved at little inconvenience.

TABLE 6-4

Typical Monthly Electrical Use in an American Home

Item	Estimated kwh per Month
Dishwasher	30
Range with oven	100
Toaster	3
Waste Disposal	3
Freezer	100 to 150
Refrigerator	100 to 150
Clothes Dryer	83
Washing Machine	9
Water Heater	415
Air Conditioner	116
Dehumidifier	32
Whole-House Fan	25
Portable Fan	4
Window Fan	15
Lights	75
Stereo	10
Television (color)	42
Vacuum Cleaner	4

While it is beyond the scope of this book to discuss the suitability of various alternative energy sources, such as solar hot water heating, wind-generated electricity, and photovoltaics (a semiconductive device that when struck by light, converts light energy to electrical energy), they can be extremely valuable options. Some earth shelter owners have become completely independent of all outside utilities by incorporating these technologies into their designs. Guidelines on power requirements for alternative energy systems were given by T. Paul in an article in *Solar Age* (see bibliography).

Maintenance Activities

Ordinarily, earth shelter maintenance should be low, thanks to the earth covering and durable construction materials. Conventional homeowners periodically must repair roofs, gutters, and downspouts, paint sidings, and repair rotten or termite-damaged wood. In an earth shelter, most of this is eliminated, although leaks or structural problems may require repair work. Most earth sheltered structures will have some minor leaks. As long as the sources can be identified, this is not a serious repair problem. Usually some localized excavation and the application of an appropriate waterproofing material such as bentonite will suffice. Only when numerous and widespread leaks occur or when the source of leaks cannot be found will the situation become serious.

As stated in chapter 4, repairs made from the inside by injecting epoxy or bentonite have proven to be effective and much less expensive than outside repairs. However, if the earth shelter is not made of poured concrete, inside repair may be impossible, and the only alternative

may be repair from the outside. Structural repairs are almost always expensive and generally involve the hiring of professionals, the use of powerful jacks for raising roofs or moving walls, extensive excavation, and the purchase and installation of any needed beams, columns, concrete, or carpentry.

Insurance and Taxes

In an article by K. Vadnais in *Earth Shelter Living*, several insurance companies were positive about covering earth shelters and are now offering substantial discounts. They are beginning to recognize the lower risks associated with well-designed and well-constructed earth shelters. They do insist that the house be designed by an architect and the plans be approved by an engineer. Most companies also want assurance of strict code adherence. Specific items usually examined by insurance companies include contractor experience and ability, electrical wiring, air quality, structural integrity, liability, soil conditions, marketability, and dwelling costs.

Insurance companies will more meticulously examine an owner-built home than one that has professional involvement. However, if the home is well built, there should be no problems in securing coverage. Although a concrete structure and bentonite waterproofing are preferred by most of the companies interviewed, all indicated that they would consider covering any home with an approved structural or waterproofing system.

Although providing detailed guidance on tax matters is beyond the scope of this book, it is clear that some conservation and energy aspects of earth sheltered homes may qualify for federal and state tax credits. Trombe walls, direct-gain components, and other passive solar features may be eligible, and earth shelter owners should keep specific data on costs and construction of such items for tax purposes. Some states provide property tax refunds or other incentives for energy-efficient homes. Tax guides and information sources (available from the Internal Revenue Service, university extension of-

---

TABLE 6-5

Energy-Saving Practices

— purchase only energy-efficient appliances (refer to EnergyGuide labels, supplied by manufacturer indicating predicted average energy use.)

— locate freezers and refrigerators in cool areas with good air circulation and away from heat sources

— replace old and worn gaskets on freezers and refrigerators

— match the size of pots and pans to the size of the burner on a stove, and cover pans while food is cooking

— minimize the amount of water in cooking pans

— wash only full loads in dishwashers, clothes washers and dryers, and be sure the lint screen on the clothes dryer is clean

— install a timer on the water heater, and insulate all hot water pipes

— use clothes lines to dry clothes whenever possible

— turn off all lights and appliances when not in use

— install automatic door closers especially if door closing by children is a problem

— use microwave ovens, crock pots, small broilers, and pressure cookers whenever possible

— substitute one large bulb for many small bulbs for lighting, and use fluorescent lighting, which is even more energy efficient than incandescent, whatever size bulb

fices, and many banks and post offices) should be consulted.

EXPERIENCES

Temperatures, humidity levels, and indoor air quality have generally not been serious problems, and most owners report maintaining comfortable homes with comparative low energy use (see chap. 2) and a reasonable amount of involvement in the operation of the home's energy systems.

Experiences with Indoor Air Quality

Indoor air pollution has elicited a great deal of concern in recent years as air fresheners, solvents, adhesives, fire retardants, and other modern chemicals have been added to the list of products classified as health hazards in tight, energy-efficient buildings. More and more cases of adverse health reactions to interior environments are being reported (see table 6-1). However, surprisingly little information is available on the levels of indoor pollutants in homes as a class, and almost no studies have been published specifically on earth shelters, (with the exception of articles dealing with radon, to be discussed later).

In newly constructed public buildings (not earth sheltered) the following cases have been reported.

— In a new office building in San Francisco 250 out of 800 employees complained of illnesses that were attributed to fumes from cleaning and maintenance fluids and organic chemicals commonly found in human habitations. The air-handling system did not ventilate these fumes adequately.

— In West Germany, a new building has been shut down in an effort to "bake" out chemical toxins that were found in wall panels, floor surfaces, and soft vinyl fittings.

— Several reported deaths of older people have been attributed to formaldehyde poisoning in airtight interior environments.

It has recently been reported that homes with gas stoves have a greater number of complaints of children having bronchitis, day or night coughing, chest colds, wheezing, and asthma. Although other studies found no link between these complaints and gas stoves, such stoves have been found to add higher levels of pollutants to indoor environments than electric stoves. Woodstoves produce higher levels of pollutants than oil or gas furnaces, especially with reductions in draft settings. Information from Boston studies indicates that some pollutants are three to five times greater when wood is burned than when it is not.

An energy-efficient house in California had very high levels of formaldehyde that remained elevated even with some nighttime ventilation. In other studies, concentrations of nitrogen dioxide and suspended particulates were higher in tight homes (air changes of less than 0.6 air changes per hour). Again, airtight homes with gas appliances were found to be higher in some pollutants than those without such appliances.

Regarding radon concentrations, homes with between 0.5 and 1.0 air changes per hour had comparatively low levels of radioactive activity whereas homes with lower ventilation rates (0.2 air changes per hour) had concentrations several times higher than conventional structures.

In one study, earth sheltered homes in Illinois were found not to have significantly higher radioactivity associated with radon gas than aboveground buildings. The same study cautioned that earth shelters must be well built, otherwise problems may arise as radon diffuses through poorly sealed cracks and incorrectly designed ventilation systems. However, another study found from the admittedly scanty available data that earth sheltered dwellings like other tightly built structures may be susceptible to higher than normal radon levels. A Colorado study agreed.

Energy Management

Principles and experiences regarding energy performance have been given in other sections of this book (particularly, chaps. 2 and 4 and Appendix A). Here experiences in energy management by the occupants of earth sheltered homes will be discussed. Effective energy control in earth sheltered homes entails the regulation of window openings; the installation of air seals around windows and doors, of air locked entries and of door closers; the use of interior heat sources such as stoves, baths, refrigerators, and lights. The individual preferences of people for amounts of air movement and fresh air is also significant. The following experiences from earth shelter residents in Australia and the United States have been culled from reports.

— High interior temperatures in an earth shelter with small children were primarily caused by excessive air intrusion from the outside through the children's repeated use of the front door.

— An earth-covered home in Maine was overheating when the wood-burning stove was used excessively on nights prior to a sunny winter day.

— An earth-covered home with a hot tub became too hot and humid for several days with long use of the hot tub. The problem was solved by adding a vented skylight above the unit.

— Residents in Australia indicated that unvented cooking stoves caused excessively hot and humid conditions during the summer. This problem was solved by moving the stove outdoors in hot weather. Other Australian earth shelters were able to ventilate effectively by correctly operating vertical vent pipes, exhaust systems, and windows and doors.

As discussed in chapter 2, earth shelters can be significantly more efficient in energy use than most aboveground houses, but to a large extent, the difference depends on appliance use unrelated to heating and cooling. Although documented experiences are scarce, it is known that hot water usage and the operation of refrigerators and stoves have proven to be significant. See table 6-5 for some proven energy-saving practices that have made real differences in energy consumption in earth sheltered and conventional homes.

Insurance and Taxes

In some cases, insurance companies have offered discounts up to 30 percent where policies have been adjusted for the low risks characterizing an earth shelter regarding wind, fire, and other threats. Premium rates as low as $140 a year for an average earth shelter have been quoted.

Tax credits can be a significant part of the financial savings associated with an energy-efficient home. In 1980, over 2,900 renewable energy systems in Wisconsin and over 1,500 in Michigan qualified for energy tax credits. Since then, the numbers of qualifying residences have grown dramatically. More than forty states have renewable energy credits, including tax credits, direct refunds, and property tax or sales tax exemptions.

Firsthand Experiences

The construction of our earth shelter in 1979 created much interest in our community, and hundreds of people have toured the home. Most of the comments have been positive, especially concerning the bright and open atmosphere. The only negative comments have been a matter of taste.

INTERIOR AIR QUALITY

All of our appliances are electric except for the wood-burning stove. Interior finishes consist of drywall, plaster, stained and finished oak trim and cabinetry, and carpeting. The odors emanating from these finishes, while strong initially, have lost most of the "new" smell. Ventilation is provided by an air intake tube, Trombe wall vents, wood-burning stove chimney, and two bathroom vents (see fig. 6-2). Water is supplied from a local aquifer. Although the air change rate has not been measured, we estimate it to be 0.5 to 1 air changes per hour. No tobacco smoking occurs in the house, and we use average amounts of maintenance, cleaning, and other chemicals. We have two small children under four years of age. My wife has an allergic condition in which she responds negatively to pollen and molds associated with rainy weather. Even though we spend a great amount of time at home, we have experienced no ill health effects that could be attributed specifically to the earth sheltered environment.

Occasionally, we smell smoke from the wood-burning stove, but this odor usually dissipates within an hour. Other odors (see table 6-2) have not been noticed, and no measures of indoor pollutant levels have been taken except for a radon measurement. This was done using a Type B detector exposed for ninety-two days. (Such detectors are available from Terradex Corporation, 460 North Wiget Lane, Walnut Creek, California 94598.) The result, given in average picoCuries of radon-222 per liter of air was 2.19. This value is in the medium range for indoor exposures (see table 6-6) and compares favorably with other homes studied.

TABLE 6-6

Radon Concentrations and Suggested Standards

Radon Concentrations (In picoCuries per liter—pCi/l)		
Level	Indoor	Outdoor
Low	less than 0.5 pCi/l	less than 0.3 pCi/l
Medium	0.5–4.0	0.3–1.0
High	greater than 4.0	greater than 1.0

Concentrations Considered Safe

Infiltration Characteristics	Concentrations
Conventional Home	4 pCi/l
Airtight, relatively dust free	3 pCi/l
Airtight, somewhat dusty	2 pCi/l

THERMAL PERFORMANCE AND ENERGY USE

See appendix A for a detailed two-year study of the energy performance and energy use of our home. From our experience during and after this time, we have found the following to be significant.

1. Exterior rolling shutters installed on Trombe walls and windows. Temperatures and energy use have improved dramatically since installation.

2. Thermometers and relative humidity meters to determine when to open and close the house instead of subjective comfort level perception.

3. A summer exhaust fan for ventilating the home at night.

4. The summer Trombe wall vents, air intake tube, and other vents for providing ventilation. In retrospect, we would choose a ventilating skylight instead of the fixed model we now have over the kitchen.

5. Ventilation provisions for closets, baths, and other closed-in spaces, especially when the house is unoccupied for long periods.

6. Fans and overhead duct systems for moving air in both summer and winter.

7. Hot water preheat coils added to wood-burning stove and a timer to an electric hot water heater, both of which have meant energy savings. Future plans call for adding either a solar hot water heating system or a water-heating heat pump.

8. Good vegetative cover on the roof and sides of the earth shelter.

INSURANCE AND TAXES

The insurance company classified our home as a fireproof masonry building and gave us a rate that was 20 percent lower than other homes in our area. We were also able to specify the dollar amount of the expected maximum damage that we felt our home could sustain. This allows us a lower premium. The policy maximizes coverage on liability and contents and minimizes coverage for fire damage to the main structure.

Some of our passive solar features qualified for federal and state tax credits. We kept careful records on all of our expenditures and verified that our home was professionally designed. These records were needed for documenting the request for state tax credits.

RECOMMENDATIONS

For a safe and healthful environment in an earth shelter, building materials and interior furnishings that contain high levels of contaminants should be avoided. The building itself should be sealed against sources of radon gas, or the house can be ventilated at a rate that maintains low indoor pollutant concentrations. This may require the installation of heat exchangers or similar equipment. Finally, sources of contaminants (such as smoking) should be avoided.

When making decisions about ventilation, occupants should rely on objective readings from thermometers and relative humidity meters rather than subjective feelings. Excessively hot or humid air should never be introduced in an earth shelter. Rolling exterior shutters, to control solar radiation, should be included in the initial design. In summer, use small fans or other means to move air; and in winter, use wood-burning stoves but be careful to avoid overheating and air pollution.

When choosing insurance for an earth shelter, select a flexible policy that omits unneeded coverage such as wind and total fire protection. Liability for accidents around an earth shelter is of primary concern, followed by protection against vandalism, glass breakage, and damage to exterior elements.

ENERGY PERFORMANCE *

Temperatures, relative humidity levels, and energy use in a passive solar, earth sheltered home in central Kansas were monitored over a seventeen-month period. Window insulation at night had a significant effect on heat retention, and considerable heat gain and ventilation benefits were realized from two Trombe walls built into the structure. Daily inside temperatures varied from a low of 62° F in the winter to a high of 87° F in the summer, with an average room temperature of 73° F. The earth sheltered home used 30 percent less electricity than an aboveground home of similar size. The woodburning stove consumed 67 percent less wood than a conventional home, and there was no need for propane. Overall, at least 80 percent of the heating needs were supplied by passive and earth-contact strategies.

Many authors claim potential energy savings from earth-sheltering techniques, but few published studies document these savings. This paper provides data on the performance of a passive solar, earth sheltered home in central Kansas. Most earth sheltered homes rely mainly on direct solar gain, but two Trombe walls were incorporated into the passive solar design of this home, making it unique in the literature. The Trombe walls—black, glass-covered concrete structures—make up part of the south wall of the building and, with the windows, comprise the solar collection system. This combination of passive solar techniques and earth sheltering is indeed an attractive strategy, as the following discussion will show.

HOUSE DESIGN

The floor plan of the house contains 1,500 square feet of living space and 600 square feet of garage. The structure is long and narrow, which permits the winter sun to reach the back walls, and the room arrangement is open, which facilitates air movement and promotes a feeling of spaciousness (fig. A-1). The interior is light in color, and approximately 75 percent of the floor is carpeted.

The kitchen has a 30-inch skylight, while a vent pipe 10 inches in diameter runs beneath the floor slab to provide combustion air for the wood-burning stove (the only backup heat source). Each bathroom has a vent pipe opening through the roof, and each is fitted with a small exhaust fan. A ceiling fan and duct system distribute heat in the winter, and a whole-house ventilation fan can be used to pull in cool night air in the summer.

The Trombe walls, built as part of the south wall, are oriented 28 degrees east of due south; each is 8 feet high, 10 feet long, and 14 inches

* Max R. Terman. "Energy Performance of an Earth Sheltered Home with Trombe Walls." *Underground Space*, 6 (1981): 180–85. Reprinted by permission of Pergammon Press Ltd.

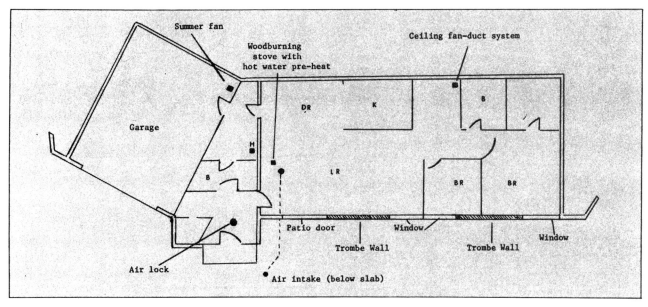

A-1. Diagram showing the floor plan and energy systems of the passive solar, earth sheltered home. (Reprinted, with permission, from *Underground Space*, Vol. 6, Max R. Terman, "Energy Performance of an Earth Sheltered Home with Trombe Walls," © 1981, Pergamon Press. Ltd.)

thick, with a mass of approximately 13,920 pounds (see fig. A-2). Inside Trombe vents are at floor and ceiling levels; outside vents for summer use open beneath a 28-inch-wide overhang. Four 4-by-6-foot windows and one 6-by-10-foot patio door are also located on the south wall and provide about 160 square feet of direct-gain collector surface. The windows are equipped with reflective blinds and heavy cloth draperies. [Note: exterior shutters have since been installed.]

The shell of the house is a poured, post-tensioned concrete structure with 9-inch-thick roofs and 8-inch thick walls, the structure being waterproofed with a combination of bentonite and premolded membranes. Six inches of polystyrene insulation cover the top of the roof over the living area (1 inch over the garage roof), and 2 inches decreasing to 1 inch cover the walls; the floor, 5-inches thick, is not insulated.

The building is covered with 3 feet of soil on the front of the roof, which grades back to 2 feet at the rear. The roof and the side and back walls are in contact with an extensive mass of earth that makes up a hill approximately three acres in size. Maximum excavation depth into the hill is 8 feet.

The hot water heater, insulated with 2 inches of Fiberglas, receives preheated water from coils in the wood-burning stove and is on a timer that turns the power on for three hours a day at periods of maximum use. The household appliances are electric.

A-2. The Trombe walls figure prominently in the facade of this passive solar, earth sheltered building. (Reprinted, with permission, from *Underground Space*, Vol. 6, Max R. Terman, "Energy Performance of an Earth Sheltered Home with Trombe Walls," © 1981, Pergamon Press, Ltd.)

PERFORMANCE DATA

Room temperatures and humidity were measured from the back living room wall with a meter for temperature and relative humidity. Trombe wall and remote temperatures were measured with an eight-probe thermistor thermometer; a mercury thermometer in a sheltered entrance took outdoor temperatures.

Temperature

Records of the thermal performance of the house date from January 1980, three months after the structure was backfilled. Due to loose soil structure and lack of vegetation, the thermal characteristics of the backfilled soil were different from conditions that can be expected after the soil settles and vegetation is established.

Monthly temperature variation, measured on the back wall, decreased from January 1980 to January 1981, with the coolest temperature (62° F) occurring in February 1980, when the house was unoccupied for eight days. During a record Kansas heat wave (over forty consecutive days with temperatures over 100° F), a high temperature of 87° F was recorded; however, no shade was yet provided by deciduous trees. The average monthly temperatures ranged from 66° F (February 1980) to 81° F (August 1980). The average internal temperature for the study period was 73° F. Only air distribution and venting fans were used during the summer, and a total of 0.8 cords of wood was burned each winter (table A-1).

The wood-burning stove and the window insulation (blinds and draperies) influenced the internal temperatures, as can be seen during two winter cold spells: January 1980 (no blinds) and December 1980 (blinds and draperies for night insulation). The use of draperies and blinds allowed the wood-burning stove to be burned at lower temperatures and still maintain a higher room temperature (fig. A-3).

The extent of the influence of the sun on house temperatures during each month can be ascertained from table A-2. During January and February 1980, the internal temperatures were 40° F higher than outside temperatures (no aux-

---

TABLE A-1

Fuel and electrical usage comparison between a well-insulated aboveground farmhouse * and a passive solar, earth sheltered home **. Winter and summer daily average outside temperatures are given in degrees F.

Date	Jan. High/low	July High/low	Average Monthly kwh/person	Cords of wood used/winter	Heating fuel (gal. propane)
Farmhouse					
6/77–12/77		89.2 70.2	300	0.5	402
1/78–12/78	24.9 8.7	89.7 68.9	409	1.8	645
1/79–6/79	20.8 4.1	84.1 67.6	313	1.2	300
(Monthly average for 2-year period)			341	1.2	449
Earth sheltered house					
1/80–12/80	36.6 20.7	97.4 72.9	244	0.5	0
1/81–3/81	40.8 19.8	86.1 70.4	234	0.3	0
(Monthly average for 2-year period)			239	0.4	0

* 1,800 square-foot living area, 2 floor levels, root cellar, 1 bath, electric hot water, forced air furnace, airtight wood-burning stove, window air conditioner, water pump. Occupied by 2 adults, both absent from home, 8:A.M.–4:00 P.M., September through May.

** Three feet of soil cover, 1,500-square-feet living space, 2 baths, 1 floor level, electric hot water heater, attached garage, 2 Trombe walls, wood-burning stove with hot water preheat coils, timer on hot water heater, duct fan, whole-house fan, water pump. Occupied by 2 adults and 2 infants continually.

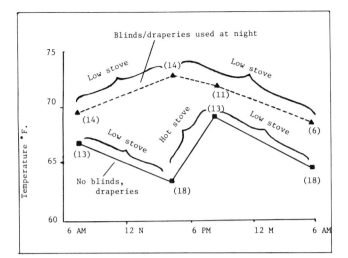

A-3. The dotted line shows the effects of using draperies. Note that the temperature stays more constant than when no night insulation (solid line) is used. Wood-burning stove use is also indicated. Note that when no window insulation is used a hot stove is required at night whereas a low stove temperature suffices with the use of night insulation. (Reprinted, with permission, from *Underground Space*, Vol. 6, Max R. Terman, "Energy Performance of an Earth Sheltered Home with Trombe Walls," © 1981, Pergamon Press, Ltd.)

iliary heat provided). During the record hot summer of 1980, internal temperatures were as much as 24° F lower than those outside, with no shade or vegetative cover for the building. This table also reflects the moderating effects of earth sheltering, as the outside extremes are not

TABLE A-2

Internal temperatures on clear days at noon during each month of 1980.

	Outside temps.	Inside temps.
Jan 19	30°	70°
Feb 17	29°	70°
Mar 17	60°	75°
April 13	58°	75°
May 17	60°	70°
June 17	84°	76°
July 15	106°	84°
Aug 18	86°	81°
Sept 14	90°	78°
Oct 19	80°	79°
Nov 19	57°	79°
Dec 14	40°	76°

reflected inside the house. (The performance of the house is expected to improve as deciduous trees and grass cover are added to the landscape.)

Table A-3 shows the mean temperatures at various locations in the house during the winter (January–March) and summer (June–July). The temperatures of the entryway, an air lock isolating the living room from the outside, show the effects of outside contact and isolation from the Trombe walls. The walls, floor slab, and ceiling temperatures reflect the earth temperatures adjacent to them; the skylight, as indicated by the wide temperature swings in this structure, is a source of heat loss during the winter.

TABLE A-3

Average temperatures for January–March 1980 and June–July 1980 for various locations in the house

	Living room	Entry-way	Sky-light	Front floor slab	Back wall	Ceiling
Jan–March	69°	51°	42°	64°	65°	65°
June–July	79°	82°	89°	80°	81°	81°

The high and low temperatures on the outer and inner surfaces indicate that the Trombe wall is an effective heat source during the winter, but may cause unwanted heat gain during the summer unless it is shaded or shuttered (table A-4; figs. A-4 and A-5). The inner surface temperature of the Trombe wall remained constant compared with the outer surface temperature. During a winter cold period (9–11 February 1980) outer surface temperatures varied by 44° F; the inner-wall range was only 7° F. This constant radiant temperature contributes significantly to the maintenance of comfortable interior temperatures. During a summer heat wave the outer surface varied by only 15° F and the interior Trombe wall surface by less than 5° F. The room temperature averaged 84° F during this time.

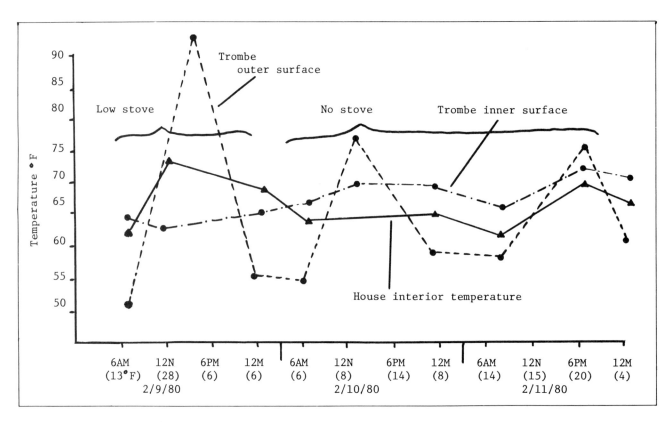

A-4. Thermal performance of Trombe walls and house interior during winter (February 9–11, 1980). Outside temperatures in parentheses below the time of day. (Reprinted, with permission, from *Underground Space*, Vol. 6, Max R. Terman, "Energy Performance of an Earth Sheltered Home with Trombe Walls," © 1981, Pergamon Press, Ltd.)

A-5. Thermal performance of Trombe walls and house interior during a summer heat wave (July 8–10, 1980). Outside temperatures below the time of day. (Reprinted, with permission, from *Underground Space*, Vol. 6, Max R. Terman, "Energy Performance of an Earth Sheltered Home with Trombe Walls," © 1981, Pergamon Press, Ltd.)

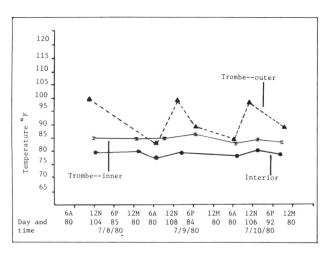

TABLE A-4

Temperatures recorded on Trombe wall surfaces during summer and winter periods, 1980.

	Highest temp. °F	Lowest temp. °F	Average at noon °F
Outer Trombe Surface			
Jan–March	101°	50°	86°
June–July	104°	78°	91°
Inner Trombe Surface			
Jan.–March	76°	59°	68°
June–July	89°	78°	84°

Relative Humidity

Figure A-6 shows the monthly variations in inside temperature and humidity levels, highest during April, May, and June (64–66 percent) and lowest during the winter months (33–39 percent). The first year, 1980, had higher humidity levels than 1981, presumably because moisture was released from the structure's curing concrete.

The average relative humidity dropped 10 percent (from 60 to 50 percent with the opening of the outside (summer) Trombe wall vents (fig. A-7). (The bottom 10 percent of the surface of the Trombe walls is exposed to the sun to

A-6. Room temperatures and humidity levels (H) by month (average outside temperatures below the month). (Reprinted, with permission, from *Underground Space*, Vol. 6, Max R. Terman, "Energy Performance of an Earth Sheltered Home with Trombe Walls," © 1981, Pergamon Press, Ltd.)

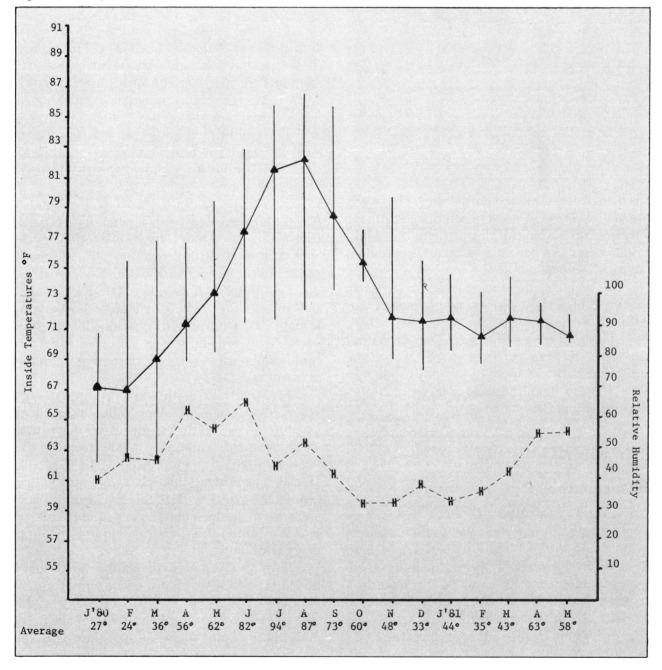

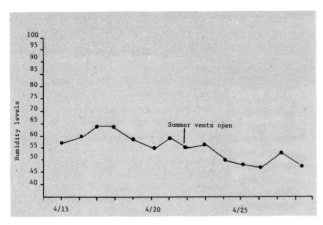

A-7. Room humidity fluctuations and summer vent operations. Average outside humidity was 85 percent during this period. (Reprinted, with permission, from *Underground Space*, Vol. 6, Max R. Terman, "Energy Performance of an Earth Sheltered Home with Trombe Walls," © 1981, Pergamon Press, Ltd.)

promote this summer ventilation.) The walls showed no signs of condensation, even during the most humid months; the skylight dome and window surfaces had small amounts of condensation.

The fuel use and electrical use for the earth sheltered home are shown in table A-1. For comparative purposes, data from a well-insulated farmhouse of similar size for the period June 1977 to June 1979 are given. More time during the day was spent in the earth sheltered home than in the farmhouse, which was unoccupied during working hours.

DISCUSSION

Temperature

Several observations by other researchers were repeated in our data. Due to the effects of radiant heat, optimal humidity levels, and good ventilation, the temperatures in the house were comfortable, even at 65° F or 84° F. According to one researcher, mean radiant temperatures provide comfortable conditions at relatively high or low air temperatures; in our study this was true even at 87° F. The earth-backed walls and the large thermal mass of the building facilitated a constant, slowly changing temperature,

which added to the maintenance of comfortable conditions.

The lowest temperature (62° F), reached in February 1980, agrees with recorded earth temperatures for similarly designed structures in temperate and cool climates. Apparently, this temperature is a winter baseline value, the result of the earth-contact factor and minimum heat gains.

Thermal performance of earth sheltered buildings can generally be expected to improve as the soil settles and vegetation grows. The temperature swings of the building for January–May 1981 were more moderate than those for 1980. This may have been due to different weather conditions, however; tests need to be run on earth sheltered buildings of different ages in the same geographical area to substantiate this claim.

Insulation placement is very important to the performance of earth sheltered structures. The 6 inches of polystyrene insulation on the roof of this structure, together with the earth cover, gives the roof an effective R-value of 25–30 and the walls a rating of R18–R22, equivalent to some of the best-insulated aboveground structures. The low infiltration losses from earth sheltered buildings complement these high insulation values, to produce a thermally efficient building.

With night insulation and minimal use of a woodstove, it was possible to maintain a temperature of 70° F during the coldest weather. Without such insulation, it was necessary to burn substantially more wood (see fig. A-4). During the summer, the reflective blinds helped reduce the window heat gains. In a building of such high thermal mass, the judicious use of window blinds and draperies can improve thermal performance.

Careful placement of ceiling and whole-house fans to distribute heat in winter and to ventilate in summer is also important, as the ability to warm or cool the thermal mass (by drawing in cooler night air) can make a difference the following day. Once a temperature is reached, it changes by only 3°–5° F over twenty-four hours (see fig. A-3).

The Trombe walls can facilitate both winter heating and summer ventilation. By opening the heating (inside) vents on a cold, sunny day, it is possible to achieve temperatures above 70° F, even on the coldest winter days (see tables A-2 and A-4). During the summer, the operation of summer (outside) vents can indirectly affect comfort levels by influencing air movement and humidity levels (see fig. A-7). It is calculated that passive solar techniques in Kansas can supply up to 80 percent of the heating needs in a well-insulated aboveground building. In this house at least 80 percent of the heating needs were met by passive and earth-contact strategies (see table A-1). It is estimated that a 50-percent savings can be realized with the earth-sheltering techniques (excluding Trombe walls).

An often overlooked advantage of earth sheltering is the summer cooling provided by earth contact. In the house under study, the floor slab was not insulated and the painted interior walls were exposed. This strategy proved effective for the Kansas climate; summer house temperatures were as much as 24° F below those outside. The floor slab was not uncomfortably cold in the winter (64° F). Moreover, summer performance should improve. Cooling by earth and vegetation can be significant, according to several researchers; one estimates that a 1,200-square-foot sod-covered roof can provide a cooling potential of 1.5 million Btu per day.

Relative Humidity

Moisture control is a major concern in the operation of an earth sheltered building. In this house the humidity levels were controlled by the Trombe walls, stove and bath vents, ceiling fans, whole-house fans, and insulation placement. The levels were comfortable even though the concrete was still losing water in the curing process. The Trombe walls, considered to be primarily heating sources, proved their value in ventilation and humidity control by promoting a slow mass movement of air, especially during the summer. The chimney and bath vents could be opened to provide ventilation powered by the wind or fans. The ceiling fan facilitated internal air movement, and the whole-house fan was used to pull in cooler night air and to vent odors. The placement of insulation on the outside of the walls (2 inches decreasing to 1 inch at the base) prevented condensation on the internal wall surfaces.

Temperature and humidity interact to provide comfort but vary with climate, so that strategies must be implemented in accordance with climatic zone. For conditions in south central Kansas, the above-mentioned strategies were appropriate.

Energy Use and Related Benefits

An earth sheltered home such as the one under investigation can save significant amounts of money. For this house, electrical bills were 30 percent less, wood-burning costs 67 percent less, and fuel (propane) costs 100 percent less than for a comparable aboveground structure. These reductions in energy usage stemmed from less heating, cooling, water heating, and lighting, and were accompanied by increased comfort—radiant heat, instead of forced air, and a "heated" garage, for example. To these must be added the environmental advantages from reduced energy use and minimal land destruction.

CONCLUSION

A passive solar, earth sheltered home in central Kansas was monitored and analyzed with respect to thermal performance, relative humidity levels, and energy use. A combination of passive solar techniques (Trombe walls, direct gain) and earth contact proved to be effective strategies for providing comfortable living conditions (temperatures of 65°–84° F and relative humidity of 30 to 60 percent) with substantial energy savings.

PROFESSIONAL HELP

No earth sheltered home should be built without the advice and counsel of people knowledgeable in the field. This section, while not exhaustive, lists architects, builders, designers, and other experts in the earth shelter movement. Arranged by state and city, the list is not an endorsement for any of the individuals or firms. Each will have to be contacted and then evaluated by the potential owner of an earth sheltered home.

Potential owners are encouraged to consult first with educational institutions involved in earth shelter research, such as the Underground Space Center (University of Minnesota) and the School of Architecture (Oklahoma State University), in their decision-making process. Other institutions, such as Clemson University,

University of Southern California, Texas Tech University, Trinity University (Texas), University of Texas at Arlington, Kansas State University, University of Missouri, Arizona State University, University of Arizona, Massachusetts Institute of Technology, University of New Mexico, University of Oregon, Washington State University, University of Washington, and the University of Wisconsin, are involved in earth shelter research and can provide a good source of unbiased information. For a more complete list of personnel associated with the American Underground Space Association, write to the Underground Space Center, 790 Civil and Mineral Engineering Building, 500 Pillsbury Drive S.E., Minneapolis, Minnesota 55455.

	ARCHITECT/ EXPERT	DESIGNER/ BUILDER		ARCHITECT/ EXPERT	DESIGNER/ BUILDER
Alabama	J. Frank Burford Birmingham	Cowin and Company Birmingham	Idaho	James W. Chase Pocatello	Scott Earth Homes Coeur d'Alene
	Giattina and Partners Birmingham		Illinois	American Colloid Co. Skokie	American Solatron Centralia
Arizona	Post-Tensioning Inst. Phoenix	Concept 2000, Inc. Phoenix		Sealant and Waterproofers Inst. Glenview	Davis Caves Chicago
	James Scalize Tempe	Earth Systems Phoenix		Tigerman and Assoc. Chicago	U-Bahn Earth Homes Granite City
California	Ralph G. Allen Santa Ana	Earth Sheltered Structures Pinole		Portland Cement Association Skokie	
	Roland Coate Venice	The Reinforced Earth Company Sacramento		Prestressed Concrete Inst. Chicago	
	David Wright Sea Ranch			Richardson, Severns, Scheeler, Greene, and Assoc. Champaign	
Colorado	Charles A. Lane Denver	Colorado Sunworks Boulder			
	Brian Larson Denver	Tecton Corporation Colorado Springs	Indiana	Clyde N. Poppe La Porte	S.E.E.D. Design Warsaw
	David Beal Boulder				Earth Castle Homes Indianapolis
Connecticut	Kenneth Labs New Haven	Raymond Cahill Wallingford	Iowa	J. D. Bloodgood Des Moines	Earth Sheltered Housing of Iowa West Des Moines
	Herbert S. Newman New Haven				Earth Sheltered Housing Systems Marshalltown
	Douglas Orr New Haven				Solarglass Earthen Homes Algon
Florida	John B. Langely Winter Park	Glen Nielson Brandon			
	William Morgan Jacksonville	R. Charles Scott Sarasota	Kansas	Keith Christensen Manhattan	Doug Deeds McPherson
Georgia	Energy Efficient Environment Atlanta	J. D. Kimsey Roswell		F. Gene Ernst Manhattan	James Harms Ulyssess
				Brian Bailey Wichita	Gary Rickman Wellsville

	ARCHITECT/ EXPERT	DESIGNER/ BUILDER		ARCHITECT/ EXPERT	DESIGNER/ BUILDER
Maryland	National Ready Mixed Concrete Assoc. Silver Spring	M. S. Milliner Construction Frederick	Minnesota (cont.)	Criteria Architects St. Paul	
Massachusetts	John E. Barnard Marstons Mills			Design Consortium Minneapolis	
	Thomas Bligh Cambridge			Charles Fairhurst Minneapolis	
	Earl Flansburgh Boston			Richard R. Egan Willmar	
	Haven Eisenberg Assoc. Boston			Environmental Design Minneapolis	
	Hugh Stubbins and Assoc. Cambridge			Howell, Radloff, Thorpe Minneapolis	
	Malcolm Wells Brewster			Dennis Johnson North Branch	
Michigan	Clark and Walter Traverse City	Naturewood Homes and Domes Ypsilanti		Michael McGuire Stillwater	
				Eldon Morrison White Bear Lake	
	Natural Alternatives Northville	Systems 2000 Sterling Heights		Myers and Bennett Edina	
	E. T. Vernulen Grand Rapids	Terra Hab Co. Plymouth		William Scott Minneapolis	
		Cave Enterprises Ann Arbor		Sticks and Stones Design Minneapolis	
Minnesota	Architectural Alliance Minneapolis	Ellison Design and Const. Minneapolis		Truman Howell Minneapolis	
	Thomas Atchison Minneapolis	Everstrong Redwood Falls	Missouri	Truman Stauffer Kansas City	Betterway Underground Homes Kansas City
	Berg and Assoc. Plymouth	Natural Spaces North Branch		Nolan Augenbaugh Rolla	R. A Burnett and Associates Rogersville
	Carmody and Ellison St. Paul	Seward West Redesign Minneapolis		The Binkley Co. St. Louis	Earth Sheltered Home Designers Cape Girardeau
	Close Assoc. Minneapolis				
	Jim W. Cox Marine on St. Croix				

	ARCHITECT/ EXPERT	DESIGNER/ BUILDER		ARCHITECT/ EXPERT	DESIGNER/ BUILDER
Missouri (*cont.*)	Larry Atkinson Edgerton	Simmons and Sun High Ridge	New York (*cont.*)	Swanke, Hayden, Connell, and Partners New York City	Lynn E. Elliott Oneida
		Terra Dome Independence	Ohio	Effective Building Products Cleveland	Malcolm Kennedy Newark
Montana	John Hait Missoula	Earthhome Construction Bozeman		Joseph Kawecki Columbus	Sheltera Homes Springfield
Nebraska	Gary Nielsen Omaha	Bright Prospects Lincoln		Richard Ohanian Springfield	Solar-Earth Energy Reynoldsburg
	John Gulick Lincoln	Down-to-Earth Homes Omaha		Gerald Pierron Portsmouth	Underground Homes Portsmouth
	Robert Youngberg Lincoln	Energy Information Omaha		Dick Strayer Dublin	U. S. Systems Logan
North Dakota	Ken Johnson Fargo	Sterling Volla Clifford	Oklahoma	Lester Boyer Stillwater	Don McCarthy Tulsa
New Hampshire	Bruce Anderson Harrisville			Alan Brunken Stillwater	David Romberg Shawnee
	Don Metz Lyme			Walter Grondzik Stillwater	
New Jersey	Louis DiGeronimo Fairlawn	J. W. Burnley Ramsey		Bishop-Kneeland Norman	
	Short and Ford Princeton			Chadsey/ Architects Tulsa	
New Mexico	Ron McClure Tijeras			Everett Piland David Architects Tulsa	
	Wybe van der Meer Albuquerque			McCune, McCune, and Assoc. Tulsa	
New York	The De Wolff Partnership Fairport	Henry J. Jacoby New York City		Arlyn Orr Stillwater	
	Easton and La Roca New York City	Norman Mendenhall New York City	Pennsylvania	Herman De Jong Quakerstown	Robert J. Carlson Palmyra
	Johnson and Burge New York City	John C. Mullins Green Island		J. L. Harter Allentown	D. D. Brennan Lancaster
				Vincent Kling Philadelphia	Penn Seiple Sunbury

	ARCHITECT/ EXPERT	DESIGNER/ BUILDER		ARCHITECT/ EXPERT	DESIGNER/ BUILDER
Pennsylvania (*cont.*)	Natural Architecture Honesdale	Terry Lees Feasterville	Washington	L. A. Riley Spokane	In-Earth Design Spokane
	Shelter Design Group Stony Run			Miller/Hull Partnership Seattle	Central Premix Concrete Spokane
South Dakota		Con-Pour Construction Rapid City	West Virginia		Underground Construction Wheeling
Texas	Coffee and Crier Austin	Earth Habitats Dallas	Wisconsin	Dennis L. Ruppel Sheboygan	Earth Shelter Corporation Berlin
	Frank Moreland Fort Worth	Geobuilding Systems Hereford		Kenneth Clark Madison	Sunpower Construction Minocqua
	Jay Swayze Hereford	Terra Set Earth-Covered Homes Kerrville			Richard D. Herr Dousman
Virginia	David, Smith, Carter Reston	Architerra Arlington			Under-the-Earth Homes Cable
Vermont	Calco, Inc. St. Johnsbury	Francis Blair Waterville			Central County Builders Weyauwega

AWARD-WINNING
EARTH SHELTER DESIGNS

A FARM HOME IN CENTRAL WASHINGTON STATE *

The project (by Robert Hull, Miller/Hull Partnership, Seattle, Washington) was to design a house for a family of four on a small farm site (fig. C-1). The client expressed interest in a house that would be as self-sufficient in energy use as possible, including passive solar design. Because of the family's living and entertainment style it was requested that the design focus on a greenhouse that could function both as living areas and solar collection space.

The site is located in central Washington state in a harsh climate that is hot and dry in the summer and cold in the winter. Temperatures vary from 110° F to extended periods of freezing. Irrigated farming is the main industry in the area.

It was decided to "go underground" early in the design process. The winter heat-loss and summer heat-gain calculations were conclusive. In addition, the resultant structural system of concrete walls and slabs provides mass to the house that can be used to store the solar gain from the greenhouse and to minimize tem-

* From E. Frenette, "Earth Sheltering: The Form of Energy and the Energy of Form," *Underground Space*, 6 (1982): 374–77; reprinted with permission of Pergammon Press Ltd.

perature fluctuations. The constant earth temperature of 45° F to 50° F, along with buried culverts/ducts, will be used for summer cooling.

Functions have been skewed in plan to create a centrally focused greenhouse surrounded by the house proper (fig. C-2). The concrete floor slabs tier up from 12 feet at the south to 8 feet at the bedroom/kitchen. Entry to the house is by tunnel.

The final solution was generated because of the inherent restrictions of a greenhouse (they are fine for winter but overheat in the summer). The architects felt basically that it would be best to "take away" the greenhouse in the summer by using commonly available wood-framed, residential garage doors as the variable skin (fig. C-3). The doors function as a greenhouse wall in the winter and then roll up completely in the summer, opening the greenhouse to the outside. They then become the summer shading device for the stationary glass roof without destroying the vertical angle of view from the house. In the summer, the interior skin of the house at the bedrooms, kitchen, and family room functions as the lockable and insect-proof layer, and the greenhouse becomes an outside porch (fig. C-4). In the winter it becomes part of the enclosed interior space (fig. C-5).

This house is an attempt to incorporate en-

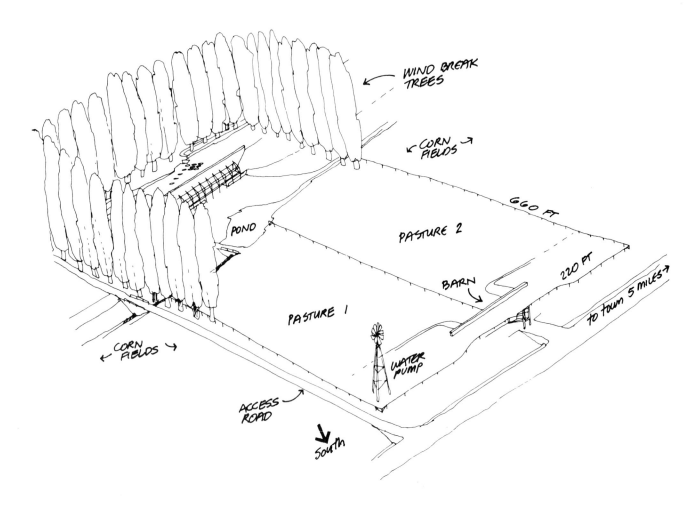

WIND BREAK TREES

CORN FIELDS →

660 FT

PASTURE 2

220 FT

POND

BARN

to town 5 miles →

PASTURE 1

← CORN FIELDS →

WATER PUMP

ACCESS ROAD

South

C-1. Overview of farm site for earth-covered home. (Reprinted, with permission, from *Underground Space*, Vol. 6, No. 6, Robert Hull, "A Farmhouse in Central Washington State," © 1981, Pergamon Press, Ltd.)

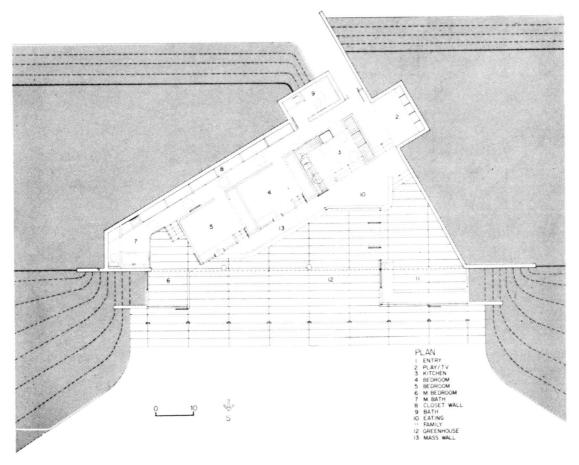

PLAN
1 ENTRY
2 PLAY/TV
3 KITCHEN
4 BEDROOM
5 BEDROOM
6 M BEDROOM
7 M BATH
8 CLOSET WALL
9 BATH
10 EATING
11 FAMILY
12 GREENHOUSE
13 MASS WALL

C-2. Floor plan of earth-covered house. (Reprinted, with permission, from *Underground Space*, Vol. 6, No. 6, Robert Hull, "A Farmhouse in Central Washington State," © 1981, Pergamon Press, Ltd.)

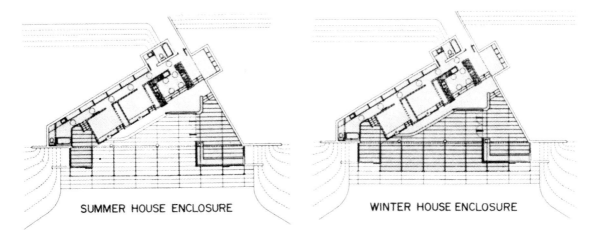

SUMMER HOUSE ENCLOSURE   WINTER HOUSE ENCLOSURE

C-3. Diagrams showing greenhouse in place in winter and "rolled-up" in summer. (Reprinted, with permission, from *Underground Space*, Vol. 6, No. 6, Robert Hull, "A Farmhouse in Central Washington State," © 1981, Pergamon Press, Ltd.)

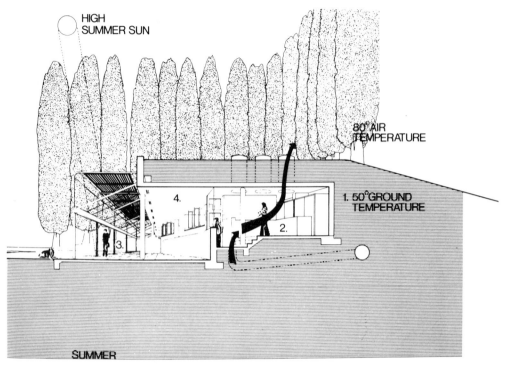

HIGH
SUMMER SUN

80° AIR
TEMPERATURE

4.

1. 50°GROUND
TEMPERATURE

3.   2.

SUMMER

C-4. Side view showing summer operation. (Reprinted, with permission, from *Underground Space,* Vol. 6, No. 6, Robert Hull, "A Farmhouse in Central Washington State," © 1981, Pergamon Press, Ltd.)

C-5. Side view showing winter operation. (Reprinted, with permission, from *Underground Space,* Vol. 6, No. 6, Robert Hull, "A Farmhouse in Central Washington State," © 1981, Pergamon Press, Ltd.)

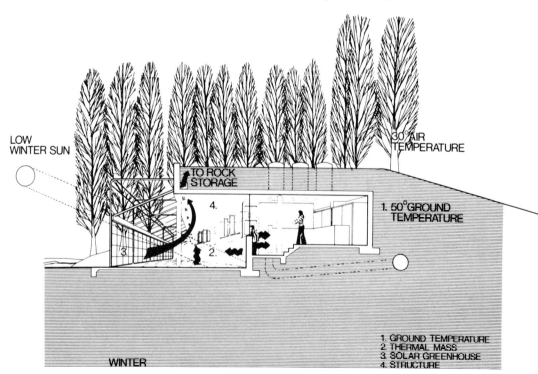

LOW
WINTER SUN

30° AIR
TEMPERATURE

TO ROCK
STORAGE

4.

1. 50°GROUND
TEMPERATURE

3.   2.

WINTER

1. GROUND TEMPERATURE
2. THERMAL MASS
3. SOLAR GREENHOUSE
4. STRUCTURE

ergy concerns into the design process and emphasizes the architect's responsibilities to a unified design approach without resorting to the clichés of solar techniques (fig. C-6).

Floors are concrete slab, etched and polished. Exterior walls are cast-in-place concrete, and interior walls are gypsum board. The roof is cast-in-place flat concrete slab with inverted beams. Lightwells are precast concrete pipe. Mechanical systems include passive solar with woodstove backup heating and radiant task heating.

C-6. Diagrams of earth-covered house with roof in place and removed. (Reprinted, with permission, from *Underground Space*, Vol. 6, No. 6, Robert Hull, "A Farmhouse in Central Washington State," © 1981, Pergamon Press, Ltd.)

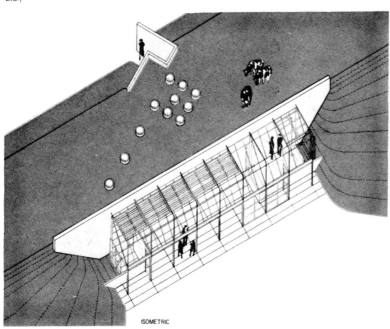

ISOMETRIC

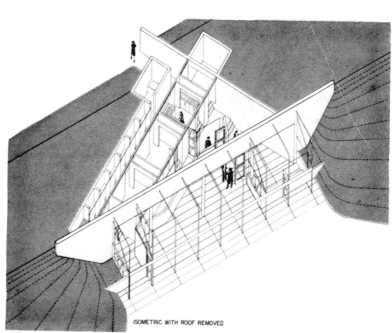

ISOMETRIC WITH ROOF REMOVED

YEAR-ROUND ENERGY MISER *

What would happen to the temperature in your house if you left it unheated from October to March?

Unless you live in the Sun Belt, chances are it would get pretty chilly inside. But a house in Frederick, Maryland, was unoccupied and unheated all last winter. "For five weeks the outside temperature dropped into the teens at night and rose into the twenties during the day," says Michael Milliner, president of the company that built the house (fig. C-7). "The lowest it got inside was 56° F, the average was 60° F, and the most it varied in twenty-four hours was 2° F."

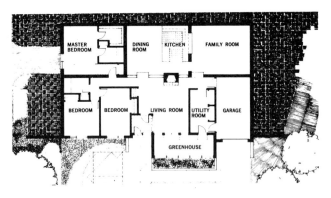

C-7. Floor plan of earth-covered house. (Reprinted, with permission, from *Underground Space*, Vol. 6, No. 6, Robert Hull, "A Farmhouse in Central Washington State," © 1981, Pergamon Press, Ltd.)

Obviously, this is no ordinary house. It is snugged into a south-facing hillside and blanketed with 18 inches of earth, which shelters it from the temperature extremes of the air (fig. C-8). Most of the windows face south and serve as passive solar collectors (fig. C-9). Company calculations indicate that the sun will supply 75 percent of the already low heating needs of the 2,300-square-foot house. The remainder, only 1.3 Btu per degree day per square foot,

* By V. Elaine Smay, *Popular Science*, August 1981, pp. 64–65; reprinted with permission of Times Mirror Magazines, Inc.

compares with 7 to 10 Btu per degree day per square foot for the average modern house. "Burn less than a cord of hardwood in the airtight woodstove, and you've got it," says Milliner.

The alternative backup heat source is a heat pump, which can also supply air conditioning —if needed. It may not be. A third of the cooling load will be supplied by a heat-pump water heater. Passive cooling is also part of the design (fig. C-10). An 80-foot-long, 12-inch-diameter concrete pipe is buried at an average depth of 10 feet on the north side of the house. Air drawn in through the pipe will average about 15° below ambient temperature, calculations indicate.

The greenhouse and clerestory windows consist of a four-layer sandwich that "could revolutionize passive solar heating," says Milliner with enthusiasm. The two outside layers are glass, but the inner layers are a new film developed by 3M. The sandwich adds up to an R-value of 3.9 (compared with R-1.88 for standard double-glazed windows). "These windows outperform double-glazed windows that are exposed eight hours a day and covered with R-6 insulation for sixteen hours," Milliner says.

The walls of the house are of 12-inch reinforced concrete block, and the roof is made of hollow-core precast-concrete planks. The house is waterproofed with bentonite, an expansive clay, and is heavily insulated on the outside, which lets the concrete serve as thermal mass to store heat and even out diurnal temperature swings.

The house won both a design award and a construction grant in the HUD/DOE Cycle 5 Residential Solar Demonstration Program. Robert May was the architectural consultant, Michael Tallmon was the solar designer, John Darnell served as a consultant, and Robert Whitesell was the structural engineer. M. S. Milliner Construction, Inc., is asking $144,800 for the house and its 3½-acre wooded lot. "People are going to have to bite the bullet up front for a passive solar, earth sheltered house that is properly built," Milliner says. "But they will not only save energy. This house requires vir-

C-8. On a winter day, the sun shines through the clerestory windows, providing direct solar gain. (Reprinted, with permission, from *Popular Science* © 1981, Times Mirror Magazine, Inc.)

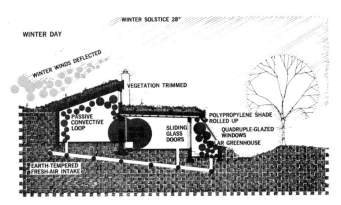

C-9. The greenhouse can provide either direct or isolated solar gain. For direct gain, the sliding glass doors between the living room and greenhouse are left open; for indirect gain, the doors are closed. The heated air in the greenhouse rises, passes through vents to the hollow cores of the concrete roof panels, and flows into the rooms at the back of the house. The warm air gives up some of its heat to the concrete mass. Cooler air near the floor is channeled back to the greenhouse through ductwork (not shown) to complete the convective loop. If necessary, the air handler of the heat pump can be used to boost air flow. At night, the heat stored in the concrete is released to the rooms. The woodstove or heat pump can provide backup heat if needed. Earth-tempered (about 50°) makeup air, which may be needed when the woodstove is used, can be drawn in through a buried pipe (see text). (Reprinted, with permission, from *Popular Science* © 1981, Times Mirror Magazine, Inc.)

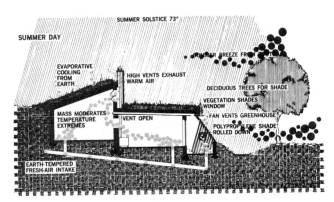

C-10. In summer, the earth pipe cools and dehumidifies incoming air, which enters the house through the ducts, rises as it warms, and exits at high vents behind the chimney. If necessary, an exhaust fan in the chimney or the air handler can be used to boost air flow. Or the heat pump can provide air conditioning. A thermostatically controlled fan exhausts hot air from the greenhouse; a polypropylene-mesh shade blocks 80 percent of the sun. The cool earth around the house acts as a heat sink. (Reprinted, with permission, from *Popular Science* © 1981, Times Mirror Magazine, Inc.)

tually no exterior maintenance, isn't going to burn, termites and rot won't touch it, and it's virtually storm-proof.''

The company (302-A East Patrick Street, Frederick, Maryland 21701) sells construction drawings for $100 a set and offers thirty slides of the design and construction, along with a synopsis of the procedures, for $30.

A TWO-STORY CONTEMPORARY HOUSE *

This contemporary, two-story house is part of a seventeen-unit development (figs. C-11 and C-12). The house is buffered from winter winds by evergreen vegetation to the northwest and from the heat of early morning and late afternoon summer sun by deciduous vegetation to the southwest and southeast. Half of its wall area is sheltered by extensive earth berms; the roof is also partially earth covered. The house is heavily insulated, even for a building in the cold Minnesota climate. The main roof of the building has an R-value of 50. Above-grade walls have an R-value of 19. The doors have a value of R-10, and windows are triple glazed.

The building's energy consumption is also minimized by an innovative floor plan and the use of buffer spaces. The living areas with different heating needs have been placed on different levels. The upper floor consists of a master bedroom, two additional bedrooms, a bathroom, and storage space (fig. C-13). The lower floor includes most of the daytime living spaces: the living room, family room, kitchen, bath, and a large atrium (fig. C-14). The garage, entry, and foyer are at an intermediate level. Each room has a separate electric resistance heating unit with individual thermostatic control. The occupants have the option of providing auxiliary heating to the lower level during the day and to the upper level at night.

The house has a well-integrated passive heating system. Three passive collection types are used: (1) over 70 square feet of triple glass on the south walls of the family room, living room, and two bedrooms provide direct heating

---

* Original design by Berg and Associates, Plymouth, Minnesota; reprinted by permission.

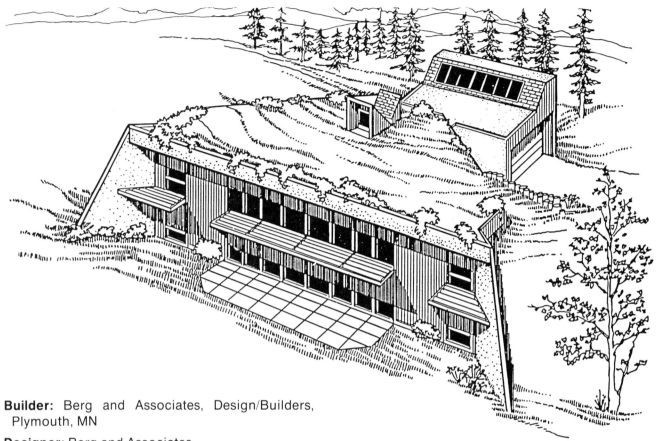

Builder: Berg and Associates, Design/Builders, Plymouth, MN

Designer: Berg and Associates

Solar Designer: Berg and Associates

Price: $120,000

Net Heated Area: 1665 ft²

Heat Load: 76.5 x 10⁶ BTU/yr

Degree Days: 8054

Auxiliary Heat: 0.99 BTU/DD/ft²

Passive Heating System(s): Direct gain, indirect gain, isolated gain

Recognition Factors: Collector(s): South-facing panels, glazing, 560 ft² Absorber(s): Concrete block wall, concrete floor Storage: Concrete block wall, concrete floor—capacity: 45,116 BTU/ °F Distribution: Radiation, natural and forced convection Controls: Movable insulation on Trombe walls, roof overhang

Back-up: Electric resistance heaters (30,000 BTU/H)

C-11. Design factors for an earth-covered house in Minnesota.
(Source: Berg and Associates, Design/Builders, Plymouth, Minnesota)

C-12. An award-winning design of a passive solar earth sheltered home by Berg and Associates. (Source: Berg and Associates, Design/Builders, Plymouth, Minnesota)

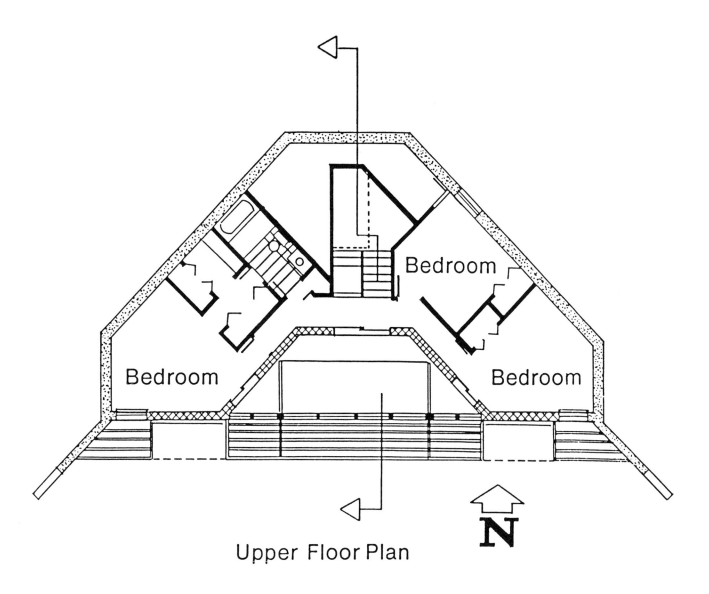

Bedroom

Bedroom

Bedroom

Upper Floor Plan

N

C-13. Upper floor plan. (Source: Berg and Associates, Design/Builders, Plymouth, Minnesota)

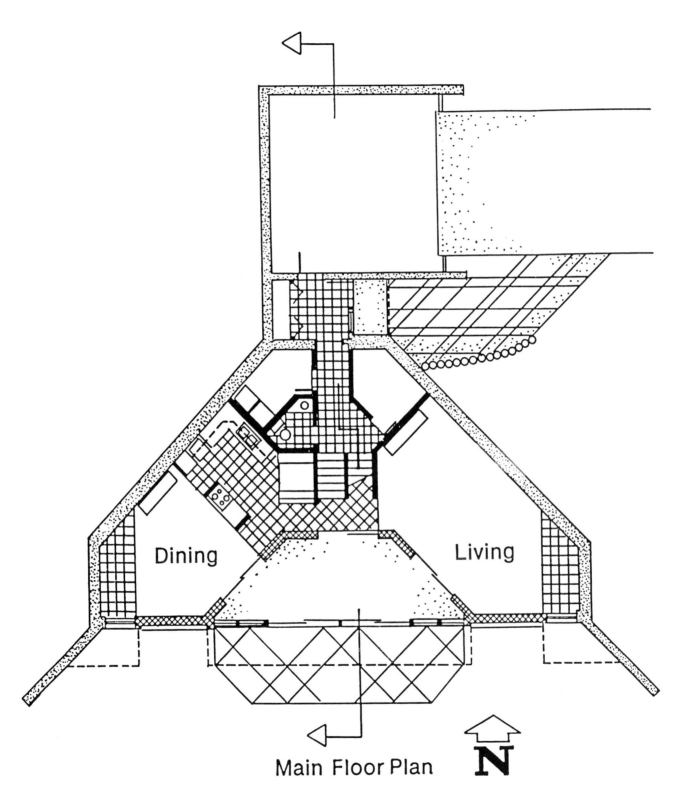

Dining

Living

Main Floor Plan

N

C-14. Main floor plan. (Source: Berg and Associates, Design/Builders, Plymouth, Minnesota)

of these rooms; (2) nearly 200 square feet of Kalwall glazing is used on two radiant Trombe walls; (3) nearly 300 square feet of triple glass on the south side of a two-story atrium heat the rest of the house. (See fig. C-15.)

Heat is absorbed and stored in the first system by a massive brick floor and a solid concrete-block wall. The other two systems absorb and store their heat in over 850 cubic feet of concrete wall and floor. Heat is distributed by natural convection and radiation, assisted by a fan located at the top of the atrium. Control of solar heat gain, heat loss, and ventilation is essential to maintaining indoor comfort. Automatic roll-down insulation on the outside of the Trombe wall controls nighttime heat loss during the winter. Baffles that deflect warm, fan-blown air from the atrium down into the center of the house during winter can also be moved to a summer position that aids ventilation. The summer position of the baffles, the open clere-

story windows, and the scoop-shaped roof above creates a venturi effect for natural suction of warm air up and out of this high area during the summer. A 4-foot-6-inch roof overhang and a similarly sized intermediate louvered overhang located over all south windows diminishes summer heat gains.

Manual dampers located above and below the Trombe glazing permit ventilation and avoid heat buildup during summer months.

The house also includes an extensive array of energy-saving appliances: water conserving bath and toilet fixtures; an energy-efficient water heater; an energy-saving refrigerator; a microwave oven; and fluorescent light fixtures. In addition, a seven-panel active solar, domestic water-heating collector array is mounted on the garage roof. This system's efficiency is enhanced by a white stone roof, in front of the collectors, that acts as a reflector.

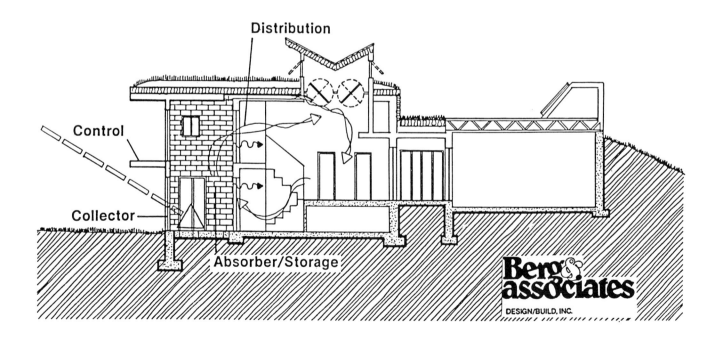

C-15. Passive solar heating system. (Source: Berg and Associates, Design/Builders, Plymouth, Minnesota)

PREDICTING COSTS FOR EARTH SHELTERED HOMES

This section lists typical construction activities involved in the building of any earth sheltered home and gives the actual costs incurred during construction of two earth-covered homes in Minnesota. Various builders specializing in the construction of earth sheltered structures also have provided general cost estimates. It should be remembered, however, that costs will vary widely according to region, contractor experience, amount of owner labor, construction technique, and inflation rate. The following figures, though dated, should provide information to the potential earth shelter owner for a more realistic financial evaluation.

CONSTRUCTION ACTIVITIES

1. Administrative, general

    site evaluation
    title search
    surveys
    blueprints
    engineering (mechanical systems)
    solar design
    permits
    soil testing
    construction insurance

2. Site utilities

    temporary power hookup
    drains and sewer systems

3. Sitework

    back hoes, bulldozers, loaders,
    trenchers, cranes, compactors
    land clearing
    drive preparation
    excavation, soil movement
    soil treatments
    well drilling
    hauling of fill materials
    backfilling
    rough and finished grade
    landscaping
    final cleanup

4. Concrete/Masonry

    compressors, finishers, saws, rebar
    cutter and bender
    power hammers, scaffolding, pumps,
    vibrators
    concrete pumps and cranes
    footings
    forms, rebar, waterstops, and so on
    structural walls
    roof materials
    floor materials
    bearing walls

5. Precast concrete, blocks, and so on

    cranes
    panels
    labor

6. Waterproofing and insulation

184

7. Electrical

underground electric line
underground telephone
light fixtures
rough-in and finished electrical
fans
garage door openers

8. Plumbing

rough-in plumbing
water line hookup
finish plumbing

9. Mechanical

rough-in heating
rough-in ventilation
rough-in fireplace and stove
finished heating, ventilation, air
conditioning

10. Carpentry/finishing

framing, drywall work
doors and windows
ceiling, floor, and wall finishes
stairways
skylights
garage door
exterior concrete and masonry work
exterior sheeting and roofing
exterior paint and stain
finish work on drives, sidewalks,
and so on
ceramic work, paneling
interior paint and stain
trim and finish work
finish flooring
touchup work
flashing, caulking, guttering

11. Special equipment

greenhouses
cabinets
appliances
mirrors
paddle fans
vacuum systems
vents and hoods
intercoms

12. Builder's overhead

ACTUAL CONSTRUCTION COSTS

Example 1 consists of a simple elevational plan, earth-covered living area (1500 square feet) and aboveground garage (800 square feet). Structure consists of poured-concrete walls with a precast concrete plank roof. The floor is woodframe over a crawlspace used as a heat delivery plenum. Overhead costs of the land and contractor are not included. Completion date was 1977.

Example 2 consists of a two-story elevational home with earth-covered living (2000 square feet) and garage areas (500 square feet). The garage opens onto the rear of the house. Walls consist of reinforced concrete block with a precast roof.

For more complete information and photographs, see the original article, from which this information is taken: "Cost and Code Study of Underground Building: A Report to the Minnesota Energy Agency," *Underground Space* 4 (1979): 119–36.

		Example 1	Example 2
1.	Administrative/general	$4,500	$8,700
2.	Site Utilities	5,650	5,803
3.	Sitework	2,000	11,204
4.	Concrete/masonry	15,300	11,190
5.	Precast concrete	4,050	6,240
5.	Waterproofing-insulation	5,100	12,722
7.	Electrical	1,400	2,590
8.	Plumbing	2,450	2,790
9.	Mechanical Systems	450	1,792
10.	Carpentry/finishing	23,900	39,235
11.	Special equipment	4,700	3,357
12.	Builder's overhead	none	3,000
	TOTAL	**$69,500**	**$99,923**

COST ESTIMATES BY BUILDERS OF EARTH SHELTERED HOUSING *

COMPANY	COST ESTIMATE
Earth Shelter Corporation (Wisconsin)	$23 per square foot for shell (includes soil borings, excavation, concrete work, insulation, waterproofing, backfilling, finish grading)
Under-the-Earth Homes (Wisconsin)	$24 to $27 per square foot for shell (includes completed exterior, completed interior, framing, plumbing, electrical)
Central County Builders (Wisconsin)	Typical house cost of $50,000 plus cost of lot
Simmons and Sun (Missouri)	$27 to $30 per square foot for shell (includes plans, engineering, excavation, concrete work, rough-in plumbing and electric, drain tile, waterproofing, insulation, backfill for post-tensioned structure)
Earth Systems (Arizona)	$15 to $20 per square foot for shell for two-story dome
Earthhome Construction (Montana)	$18 to $20 per square foot for shell (includes insulation, waterproofing, rough plumbing and electrical, front wall framing, backfill)
Everstrong (Minnesota)	$26 per square foot for shell of total earth-covered home with wood walls and ceiling

* Based on information provided by K. Vadnais, in "Affordable Housing." *Earth Shelter Living* 22 (1982): 11–14.

Note: Owners can finish off a shell of an earth-covered house for an additional $10 to $20 per square foot if they either do the work themselves or independently hire subcontractors.

active solar: a system that requires external energy to run fans and pumps.

adobe: a sun-dried clay and straw brick used for construction of high mass buildings, traditionally in the southwestern United States.

aggregate: the rock or gravel that is a component of concrete.

air changes per hour (ACH): the amount of times the air in a building is replaced in one hour.

air-lock entry: an entry way that allows entrance into a building while not permitting inside air to escape. It consists of an inside and outside set of doors, one of which is closed while the other is open.

albedo: a measure of the ability of a surface to reflect solar radiation.

ambient temperature: the surrounding temperature; usually refers to the temperature outside a house.

angle of incidence: the angle formed where the sun's rays strike a line perpendicular to a surface.

anion: an atom or group of atoms having a negative charge.

aquifer: a porous layer of geological material that contains water.

ASHRAE: abbreviation for the American Society of Heating, Refrigerating and Air Conditioning Engineers, Inc., 345 E. 47th Street, New York, New York 10017.

atrium design: a design centering around a sunken courtyard, usually built on flat sites with little view or solar exposure.

attached sunspace: an attached space such as a greenhouse or solarium that doubles as a solar collector and useful living area.

auxiliary heat: conventional heat delivered to the house to supplement solar heat.

azimuth: the angular distance, measured in degrees, between true south and a point on the horizon below the sun. A positive azimuth defines an orientation east of true south, and a negative azimuth defines an orientation west of true south.

backfilling: the process of placing earth up against or on a building after excavation and construction are completed.

barrel shell: a domed, tunnel-shaped structure that is generally constructed by spraying con-

187

crete over appropriately shaped reinforcement.

bearing angle: see azimuth.

bearing capacity of soil: the ability of the soil to support the weight of a building. Soils with good bearing capacity will not settle and move beneath the building foundations.

bentonite: a natural clay (montmorillionite) that is used as a waterproofing material for underground applications. When in contact with water, bentonite swells, causing it to plug pores and cracks, and thus acts as a barrier against water.

berm: a man-made mound, embankment, or hill of earth.

base temperature: a fixed temperature used in the calculation of heating degree days and cooling degree days, usually 65° F or 78° F.

bitumens: waterproofing materials consisting of rubberized asphalt that usually come in rolls 3 to 4 feet wide. The strips of rubberized asphalt overlap and adhere to one another.

Btu (British Thermal Unit): the quantity of heat required to raise the temperature of one pound of water 1° F. One Btu equals 252 calories, approximately equal to the heat given off by burning one kitchen match.

calorie: the amount of heat needed to raise the temperature of one gram of water 1° C is a small calorie. The quantity of heat necessary to raise the temperature of a kilogram of water 1° C is a large calorie (capital C).

camber: the upward bow or bend in a roof slab. Earth-covered roofs are usually cambered so that they will flatten when the earth is placed on them during backfilling.

cantilever: a large, projecting bracket or beam that is fastened at one end only.

chimney effect: the ventilating effect generated by the exiting of rising heated air from a building. The rising air creates a vacuum that draws cooler outdoor air in through lower windows and openings.

clerestory window: an overhead window that is commonly used to illuminate the north rooms of a building.

cold joint: a joint between two concrete pours. A cold joint results when a fresh pour is placed next to a concrete pour that has already set.

condensation: act of changing a gas or vapor to a liquid. Water vapor in a room with cool walls may condense on the wall surfaces and make them wet and subject to mold and mildew growth.

conduction: the process by which heat is transferred through a static medium such as concrete by the transferring of energy from one particle to another. Concrete exposed to cool outside conditions will conduct heat from the inside of a building to the outside, the basis for earth contact cooling.

convection: the transfer of heat between a moving fluid (liquid or gas) and a surface. Heat flowing out of the vents of a solar collector is being moved by convection.

convective loop: a closed system in which hot and cold air circulate. Solar collectors that circulate air into and out of homes use convective loops.

dampproofing products: materials such as simple, easily applied asphalt coatings that are designed to prevent dampness but do not offer full protection from water that may enter a structure.

daylighting: the process of introducing natural light into multiple areas of a house. It is important to introduce adequate light and contrast while avoiding glare. Skylights and clerestories are often used to daylight earth-covered homes.

deadman anchorage: the stabilization of structures such as retaining walls by the attachment of cables with buried weights.

deciduous trees: trees that lose their leaves in autumn. Evergreens or conifers such as pines retain their leaves throughout the year. Only deciduous trees should be planted in front of south-facing solar windows.

declination: in solar applications the deviation

between true north and magnetic north that varies with different geographic areas. Compasses must be corrected for this deviation before they are used to determine directions.

degree-day (dd), cooling: a measure of the climatic cooling requirement calculated by subtracting the daily average outdoor temperatures for a region from a base temperature of 75° F or 78° F and summing the differences for a year. For example, a day when the average temperature is 90° F would contribute fifteen degree-days (90° F minus 75° F) to the total annual cooling degree-days.

degree-day (dd), heating: a measure of the climatic heating requirement calculated by subtracting the daily average outdoor temperatures for a region from a base temperature of 65° F and summing the differences for a year. For example, a day when the average temperature is 37° F would contribute twenty-eight degree-days to the total annual heating degree-days.

dehumidification: the removal of water vapor from the air; usually performed by air conditioners or dehumidifiers.

diffuse solar radiation: the component of solar radiation that has been scattered by atmospheric molecules and particles, such as sunlight on a cloudy day.

direct gain: the transmission of sunlight directly into the space to be heated. Solar radiation passing through a window and heating a room is an example of direct gain.

double-envelope house: a passively heated home that incorporates convection circulation between an inner and outer skin.

drainage systems: pipes, gravel layers, or fabrics that promote the free movement of water away from the walls of a building. Drainage systems prevent the buildup of water pressure against a wall that may then force water through cracks and pores and cause leaks.

earth-bermed house: an earth sheltered house with a conventional roof and earth mounded against the walls.

earth coupling: the placing of a floor, wall, ceiling, or other structure in contact with the soil to promote the flow of heat between the earth and the building.

earth-covered house: an earth sheltered house with soil on the roof.

earth sheltering: the deliberate use of a mass of the earth placed in contact with a structure to benefit the environment of a habitable space. The benefits may be ecological, aesthetic, economic, and/or related to land use.

earth tubes: long underground tubes of approximately 4 to 12 inches in diameter through which air can enter a house. In theory, the air is tempered by the relatively constant temperature of the earth so that intake air is warmed in winter and cooled in summer by the earth around the tube.

ecology: the study of organisms or groups of organisms and their relationship to the environment. The use of energy and the recycling of minerals are important topics in the study of ecology.

ecosystem: a system of plants, animals, decomposers, soil, water, and other physical factors through which energy flows and minerals are cycled. A major goal for humankind is to preserve the structured interrelationships of these ecological systems.

egress: an exit from a building; especially important in complying with building codes.

elevational house: an earth sheltered house that is typically covered with earth and bermed on three sides with one exposed side facing south.

energy: the capacity for doing work. Taking such forms as chemical, electrical, mechanical, or thermal, energy is commonly measured in kilowatt hours (kwh), British thermal units (Btu), joules (j), or calories (cal).

EPDM (ethylene propylene diene monomer): a resistant synthetic rubber often used to waterproof underground buildings.

equinox: either of the two times during a year

(September 22 and March 22) when the length of day and night are approximately equal.

evaporative cooling: cooling provided by water that is evaporating. The water evaporates and removes the latent heat of evaporation from the air, thus lowering the air temperature.

evapotranspiration: the process by which plant materials dissipate heat by evaporating water. Incoming solar radiation is thus dissipated and does not warm the ground or the structure.

free-span roof: a roof that spans wall to wall without interior columns and pillars.

fascia: a horizontal piece (as a board) covering the joint between the top of a wall and the projecting eaves. The mansard roofs of many earth shelters are examples.

flashing: sheet metal used in waterproofing roof valleys or hips or the angle between a chimney and a roof. In earth shelters, flashing is often used along the juncture between the top of the parapet wall and the adjacent soil.

food chain: in an ecosystem the movement of energy and nutrients from one feeding group of organisms to another in a series that begins with plants and ends with carnivores.

footing: an enlargement at the lower end of a foundation wall, pier, or column to distribute the load.

frost heave: an upthrust of ground or pavement caused by freezing of moist soil. Frost heave is especially dangerous to earth shelters along the exposed south-facing front of the house.

furred-out walls: walls that have pieces of wood nailed to the surface for attaching drywall or other interior finishing materials. Concrete walls are often furred out but should not be if they are to be used for earth coupling or passive solar heat storage.

glare: to shine with a harsh, uncomfortably brilliant light. Glare in earth shelters may be a problem near skylights and between bright windows and dark walls.

glazing: a transparent or translucent material (such as plastic or glass) used to cover windows, greenhouses, skylights, and collectors.

guardrails: a barrier or railing placed along the edge of an earth covered roof or retaining wall to reduce the likelihood of people falling off the edge.

gunite: pneumatically sprayed concrete containing small sand-sized aggregates. Shotcrete uses larger sized aggregate, similar to normal concrete.

gypsum board: an interior finish material also called plaster board, sheet rock, or dry wall.

habitat: a place where a plant or animal lives. Earth-covered houses have the potential of preserving natural habitat by retaining green, open space.

heat exchanger: a device used for exchanging polluted inside air with fresh outside air without losing the heat. In earth sheltered homes, air-to-air heat exchangers warm incoming outside air by passing it over warm interior air that is being exhausted.

heat sink: a substance that is capable of absorbing heat. Soil absorbs huge amounts of heat and is thus a heat sink for the heat flowing from an earth covered house.

honeycombing: a condition resembling a honeycomb in structure or appearance. Concrete that is not thoroughly vibrated sometimes develops honeycombing.

hybrid system: a solar system that combines elements of more than one system for collection, storage, or distribution of energy. A solar system that combines both passive and active components is a hybrid system.

hydrostatic pressure: the pressure exerted by a quantity of water on a surface. Water that pools on an earth-covered roof exerts considerable hydrostatic pressure and causes leaks.

indirect gain system: a passive solar system in which the sun first strikes a thermal mass (such as a Trombe wall or roof pond) located between the sun and an interior space. The mass absorbs the sunlight and transfers the heat to the space.

infiltration: the loss of heat from a building through

the uncontrolled exchange of air through cracks around windows, doors, walls, roofs, and floors.

infrared radiation: electromagnetic radiation with a wavelength longer than that of visible light. Infrared radiation is felt as heat.

internal heat: heat generated in a building by appliances, lights, people, or other sources not connected with the primary heating system.

isolated-gain system: a system in which heat collection and storage are accomplished by collectors that are separated from the space to be heated.

insolation: the total amount of solar radiation that has been received. Insolation includes direct, diffuse, and reflected sunlight.

insulation: materials used to prevent heat loss or gain. Small air spaces in the materials prevent heat flow by limiting conduction or convection.

insulation configuration: the pattern in which insulation is placed around an earth sheltered house.

kwh (kilowatt-hour): a unit of work or energy equal to that expended by one kilowatt in one hour (a kilowatt equals 1000 watts).

latent heat: heat given off or absorbed in a process other than a change of temperature (such as fusion or vaporization). When water evaporates latent heat is absorbed.

latitude: a distance measured in degrees north or south of the equator. The latitude of central Kansas, for example, is approximately 38.5 degrees north.

loading factors: the weights or stresses that a building must resist or support. An earth sheltered house must withstand the forces of its own weight as well as those of soil and soil expansion, backfilling, snow, rain, and other external loads that may be placed on the building.

masonry: concrete, concrete block, brick, stone, and other similar materials.

maintenance costs: costs associated with upkeep such as reroofing, painting, repairing, and the like.

mechanical systems: conventional furnaces, air conditioners, heat pumps, and other powered sources of heating and cooling.

mean radiant temperature (mrt): the average temperature of all the surfaces in a room. The mrt of an earth sheltered room is maintained by the surface temperatures of concrete floors, walls, and ceilings that are warmed or cooled by sun and earth.

methane: a colorless, odorless, flammable gaseous hydrocarbon that is a product of decomposition of organic matter and may be used as a fuel. Natural gas is methane.

microclimate: the essentially uniform local climate of a small site or habitat. The microclimate near an earth sheltered house usually is more comfortable than that farther away from the house.

monolithic pour: the process of simultaneously pouring the floor, walls, roof, and retaining walls of an earth sheltered house.

night cooling: the use of low night temperatures to cool a house.

night insulation: movable insulation used for covering solar collectors and windows at night to prevent excessive heat loss.

neoprene: a synthetic rubber of superior resistance to oils, chemicals, high temperatures, and abrasions.

parapet wall: a low wall to retain soil on the edge of an earth-covered roof.

passive solar houses: houses that depend on non-mechanical means for a substantial part of their heating and cooling. Earth shelters use the sun for heating and the earth for cooling.

penetrational earth shelter: an earth sheltered house in which the windows are intermixed with earth berms on more than one side. In a penetrational house, light, air, and view can enter the structure from several directions.

pH: the acidity or alkalinity of a substance measured on a scale of 1 to 14 with 7 representing neutrality, numbers less than 7 increasing acidity, and numbers greater than 7 increasing alkalinity.

picoCurie: a unit quantity of radiation. Radon gas is measured in picoCuries per liter.

plasticizers: chemicals added especially to rubbers and resins to impart flexibility, workability, or stretchability. Plasticizers are also added to concrete to improve workability without adding additional water.

pollution: indoor contamination of interior air from substances such as carbon monoxide, radon, dust, soot, mists, microorganisms, and formaldehyde.

polyethylene sheets: commonly available plastic sheets used in dampproofing but not recommended for waterproofing. Polyethylene is not resistant to puncturing and seams are difficult to seal.

polystyrene (expanded): insulation that consists of tiny beads fused together to form a board or panel. Commonly called beadboard.

polystyrene (extruded): insulation that consists of

closed-cell fibers. Extruded polystyrene is recommended for underground applications over expanded polystyrene. Commonly called Styrofoam.

polyurethane foam: insulation that is sprayed on round or curved structures. It is not recommended for underground applications because of its lack of resistance to moisture.

post-tensioning: the reinforcement of concrete structures by applying tension to reinforcing steel or cables after concrete has set.

radon: a radioactive gas produced by the decay of radium 226. Radon is present in many building materials such as stone, concrete, and brick and is present in the soil around a building. It may be a factor in the development of lung cancer.

rammed earth construction: a method of construction that uses compacted soil as a building material.

rebar: steel rods used as reinforcement in concrete.

relative humidity: the ratio of the amount of water vapor in the air to the maximum amount of water vapor that can be held at a given temperature.

retaining walls: reinforced walls used to hold back soil around the edges of an earth sheltered house.

retrofitting: the installation of solar apparatuses in buildings not originally designed to be solar structures.

rock storage: the use of rock to store heat collected by an active, passive, or hybrid solar energy system.

roof pond: an indirect-gain heating and cooling system that uses water on a roof as the thermal mass.

R-value: a unit of thermal resistance; the higher the R-value the greater its insulating properties.

screeding: the act of drawing a leveling device over freshly poured concrete.

setback: the required clearance or distance between a house and the road. Building-free space is required for service equipment and utilities.

shell structure: a domed or tunnel shaped house that is constructed by spraying concrete on shaped reinforcement.

shelterbelt: a barrier of trees and shrubs that protects from wind and storm.

shotcrete: sprayed concrete. Also see gunite.

slump: a measure of the stiffness of concrete. Slump tests are needed to regulate the water-to-cement ratio, which affects strength.

solar angle: the angle the sun makes with a surface.

stucco: a material usually made of portland cement, sand, and a small percentage of lime that is applied in a plastic state to form a hard exterior wall covering.

superinsulation: extra insulation added to the floor, walls, and roof of a house.

swale: a gentle, broad ditch used to divert water away from a house.

tendon: a steel cable used to reinforce a cement slab that is post-tensioned or stressed after the concrete has set.

thermal chimney: a dark, glass-covered column that uses the sun to heat air, promoting convective air currents and thus ventilating a house.

thermal lag: a slow temperature change associated with heavy mass.

thermal mass: the amount of potential heat storage capacity available in a given assembly or system. Examples of thermal mass include drum walls, adobe walls, and concrete walls and floors.

thermal nosebleed: an area that conducts heat out of an otherwise well-insulated building.

thermosiphoning: the action of rising hot air pulling in cool air at lower levels of a house. See also convection and thermal chimney.

transit: a surveying instrument used to set boundaries and determine level surfaces.

transpiration: the act of giving off or exuding watery vapor, especially from the surfaces of leaves.

Trombe wall: a masonry exterior south-facing wall, insulated from the exterior by glass, which collects and releases stored solar energy into a building by both radiant and convective means.

vapor barrier: a construction material that is impervious to the flow of moisture and air and is used to prevent condensation in walls and other insulated areas.

vapor pressure: pressure exerted by a vapor that is in equilibrium with its solid or liquid form.

ventilation: a system or means of providing fresh air.

Venturi effect: a suction or vacuum caused by a fast flowing medium such as air passing over a small opening.

water-cement ratio: the ratio of the amount of water to the amount of cement in a mix. Generally, the less water, the stronger the concrete.

waterstop: a rubber flange or other material used to seal the joint between concrete floor and walls and ceiling and walls.

waterwall: an interior wall consisting of water-filled containers (such as steel drums) that constitute a passive heating system incorporating both collection and storage.

whole-house fan: a large fan that rapidly pulls air through a house. Used in night cooling.

windbreak: a shelter from the wind. See shelterbelt.

vertical fin: a wall or projection used to protect or shade a window or opening.

BIBLIOGRAPHY

General References

Andreadaki-Chronaki, E. 1983. "Vernacular Architecture of Greece: Earth Sheltered Buildings of Santorini." In L. Boyer, ed., *Earth Shelter Protection*. Stillwater, OK: Oklahoma State University Press.

Baggs, S. 1983. "A Design Aid for Assessing the Suitability of Soils at Earth Covered Building Sites." In L. Boyer, ed., *Earth Shelter Protection*. Stillwater, OK: Oklahoma State University Press.

Baggs, J. 1983. "Management of Earth-Covered Houses for Energy Efficiency." In L. Boyer, ed., *Earth Shelter Protection*. Stillwater, OK: Oklahoma State University Press.

Baggs, D. 1983. "A Review of Insulation Materials for Australian Earth-Covered Buildings." in L. Boyer, ed., *Earth Shelter Protection*. Stillwater, OK: Oklahoma State University Press.

Boyer, L. et al. eds. 1983. *Energy-Efficient Buildings with Earth Shelter Protection*. Stillwater, OK: Oklahoma State University Press.

Boyer, L. and W. T. Grondzik. 1983a. "Comfort Assessment in Earth Covered Dwellings in the U.S." In Boyer, ed., *Earth Shelter Protection*. Stillwater, OK: Oklahoma State University Press.

Boyer, L. and W. T. Grondzik. 1983b. "Energy Performance of Earth-Covered Dwellings in the U.S." In L. Boyer, ed., *Earth Shelter Protection*. Stillwater, OK: Oklahoma State University Press.

Boyer, L., W. T. Grondzik and T. L. Johnson. 1981.

"Comfort Analysis of Earth Shelter Interiors." In L. Boyer, ed. *Earth Shelter Performance and Evaluation Proceedings*. Stillwater, OK: Oklahoma State University Press.

Grondzik, W. T. and L. Boyer. 1983. "Earth Shelter Activity in the U.S." in L. Boyer ed., *Earth Shelter Protection*. Stillwater, OK: Oklahoma State University Press.

Chapter 1

Council on Environmental Quality. 1980. *The Global 2000 Report to the President. Entering the Twenty-First Century*. Vols. 1, 2, and 3. Washington, D.C.: U.S. Government Printing Office.

Deudney, D. and C. Flavin. 1983. *Renewable Energy: The Power to Choose*. New York: Norton.

Ehrlich, P. 1974. *The End of Affluence*. New York: Ballantine.

Flavin, C. 1980. *Energy and Architecture: The Solar and Conservation Potential*. Washington, D.C.: Worldwatch Institute.

Forman, R. and M. Godron. 1981. "Patches and Structural Components for a Landscape Ecology." *Bioscience*, vol. 31, no. 10, 733–740.

Hayes, D. 1976. *Nuclear Power: The Fifth Horseman*. Washington, D.C.: Worldwatch Institute.

Hayes, D. 1977. *Rays of Hope: The Transition to a Post-Petroleum World*. New York: Norton.

194

Hayes, E. T. 1979. "Energy Resources Available to the United States, 1985 to 2000." *Science*, vol. 203, 233–239.

Lovins, A. B. 1977. *Soft Energy Paths*. Cambridge, MA: Ballinger.

Lovins, A. B. and L. H. Lovins. 1982. *Energy Unbound: Your Invitation to Energy Abundance*. San Francisco, CA: Friends of the Earth.

Miller, G. T. 1982. *Living in the Environment*. Belmont, CA: Wadsworth.

Moreland, F. L., ed. 1975. *Alternatives in Energy Conservation: The Use of Earth Covered Buildings*. Washington, D.C.: U.S. Government Printing Office.

Moreland, F. L., ed. 1979. *Earth Covered Buildings and Settlements*. Washington, D.C.: U.S. Government Printing Office.

Myers, N. 1983. *A Wealth of Wild Species*. Boulder, CO: Westview Press.

Nash, Hugh, ed. 1979. *The Energy Controversy: Soft Path Questions and Answers*. San Francisco, CA: Friends of the Earth.

Noss, R. 1983. "A Regional Landscape Approach to Maintain Diversity." *Bioscience*, vol. 33, no. 11, 700–706.

Odum, E. P. 1983. *Basic Ecology*. Philadelphia, PA: Saunders.

Solar Energy Research Institute. 1981. *A New Prosperity: Building a Sustainable Energy Future*. Andover, MA: Brick House.

Stein, R. G. 1977. *Architecture and Energy*. New York: Anchor Books.

Stobaugh, R., and D. Yergin, eds. 1979. *Energy Future: Report of the Energy Project at the Harvard Business School*. New York: Random House.

Stokes, B. 1981. *Global Housing Prospects: The Resource Constraints*. Washington, D.C.: Worldwatch Institute.

Underground Space Center, University of Minnesota. 1982. *Earth Sheltered Residential Design Manual*. New York: Van Nostrand Reinhold.

Wells, M. B. 1974. "Environmental Impact." *Progressive Architecture*, vol. 55, no. 6, 59–63.

Wilkinson, L., ed. 1980. *Earthkeeping: Christian Stewardship of Natural Resources*. Grand Rapids, MI: Eerdmans.

Woodwell, G. M. 1974. "Success, Succession and Adam Smith." *Bioscience*, vol. 24, no. 2, 81–87.

Chapter 2

A.I.A. Research Corporation. 1979. *A Survey of Passive Solar Buildings, HUD-PDR-287(2)*. Washington, D.C.: U.S. Department of Housing and Urban Development.

A.I.A. Research Corporation. 1980. *Regional Guidelines for Building Passive Energy Conserving Homes*. Washington, D.C.: U.S. Department of Housing and Urban Development.

Best, D. 1981. "Measured Output of Two Active Systems." *Solar Age*, vol. 6, no. 7, 41–44.

Bligh, T. 1975. "A Comparison of Energy Consumption in Earth Covered vs. Nonearth-Covered Buildings." In F. Moreland, ed. *The Use of Earth-Covered Buildings*. Washington, D.C.: U.S. Government Printing Office.

Carter, D. 1982. *Build It Underground: A Guide for the Self Builder and Building Professional*. New York: Sterling Publishing Company.

Emery, A. F., D. R. Heerwagen, B. R. Johnson, and C. J. Kippenhan. 1981. "Conventional versus Earth Sheltered Housing: A Comparative Study of Construction and Operating Costs for Three Cities." In L. Boyer, ed. *Earth Shelter Performance and Evaluation*. Stillwater, OK: Oklahoma State University Press.

Golany, G. 1983. *Earth Sheltered Habitat: History, Architecture, and Urban Design*. New York: Van Nostrand Reinhold Company.

Goldberg, L. F. 1983. "Underground Space Center: Monitoring Program." *Earth Shelter Living*, no. 29, 33–35.

Grondzik, W. T., L. L. Boyer, and J. W. Zang. 1981. "Analysis of Utility Billings for 55 Earth Sheltered Projects." In L. Boyer, ed. *Earth Shelter Performance and Evaluation*. Stillwater, OK: Oklahoma State University Press.

Hylton, J. 1981. "Contrasts in Energy Design and Performance for Passive Solar vs Earth Sheltered Homes." In L. Boyer, ed. *Earth Shelter Performance and Evaluation*. Stillwater, OK: Oklahoma State University Press.

Kando, P. F. 1983. "How Builders View Passive Housing." *Solar Age*, vol. 8, no. 9, 15–17.

Kern, K., Kern, B., Mullan, J., and O. Mullan. 1982. *The Earth Sheltered Owner-Built Home*. North Fork, CA: Owner-Builder Publications.

Kiesling, E. and J. Minor. 1983. "Hazard Mitigation Through Earth Sheltering." In L. Boyer, ed. *Earth*

Shelter Protection. Stillwater, OK: Oklahoma State University Press.

Labs, K. 1975. "The Use of Earth Covered Buildings Through History." In F. Moreland, ed. *The Use of Earth Covered Buildings.* Washington, D.C.: U.S. Government Printing Office.

Labs, K. 1983. "The Underground Advantage: The Climate of Soils." In H. Wade, J. Cook, K. Labs, and S. Selkowitz, eds. *Passive Solar: Subdivisions, Windows, Underground.* New York: American Solar Energy Society.

Lewis, D. and J. Kohler. 1981. "Passive Principles: Conservation First." *Solar Age,* vol. 6, no. 9, 33–36.

Montgomery, R. H. and W. F. Miles. 1982. *The Solar Decision Book of Homes.* New York: John Wiley and Sons.

Mother Earth News, The. 1983. *Homebuilding and Shelter.* Hendersonville, NC: The Mother Earth News, Inc.

Meyer, J. and C. Sieben. 1982. "Super Saskatoon." *Solar Age,* vol. 7, no. 1, 26–32.

Oehler, M. 1978. *The $50 and Up Underground House Book.* New York: Van Nostrand Reinhold Company.

Ribot, J. C., A. H. Rosenfeld, F. Flouquet, and W. Luhrsen. 1983. "Summary of International Data on Monitored Low-Energy Houses: A Compilation and Economic Analysis." In L. Boyer, ed. *Earth Shelter Protection.* Stillwater, OK: Oklahoma State University Press.

Roy, R. L. 1979. *Underground Houses: How to Build a Low-Cost Home.* New York: Sterling Publishing Company.

Scott, R. G. 1979. *How to Build Your Own Underground Home.* Blue Ridge Summit, PA: Tab Books.

Shapira, H. B., G. A. Cristy, S. E. Brite, and M. B. Yost. 1983. "Cost and Energy Comparison Study of Above- and Below-Ground Dwellings." *Underground Space,* vol. 7, no. 6, 362-371.

Shurcliff, W. A. 1981. *Super Insulated Houses and Double Envelope Houses.* Andover, MA: Brick House.

Solar Energy Research Institute. 1981. "What Do Homeowners Think: A National Study of the Residential Solar Consumer." *Solar Age,* vol. 6, no. 4, 22–26.

Solar Energy Research Institute. 1983. "The Best Passive Heating Data Yet." *Solar Age,* vol 8, no. 7, 23–28.

Stains, L. 1980. "Double Shell Houses." *New Shelter,* vol. 1, no. 6, 72–85.

Stanford, G. 1983. "Thermal Environment of the Lithotectural Dugouts of White Cliffs—Australia." In L. Boyer, ed. *Earth Shelter Protection.* Stillwater, OK: Oklahoma State University Press.

Strickler, D. J. 1982. *Passive Solar Retrofit: How to Add Natural Heating and Cooling to Your Home.* New York: Van Nostrand Reinhold Company.

Underground Space Center, University of Minnesota. 1978. *Earth Sheltered Housing Design: Guidelines, Examples, and References.* New York: Van Nostrand Reinhold Company.

Underground Space Center, University of Minnesota. 1981. *Earth Sheltered Housing: Code, Zoning, and Financing Issues.* New York: Van Nostrand Reinhold Company.

Underground Space Center, University of Minnesota. 1981. *Earth Sheltered Homes: Plans and Designs.* New York: Van Nostrand Reinhold Company.

Underground Space Center, University of Minnesota. 1982. *Earth Sheltered Residential Design Manual.* New York: Van Nostrand Reinhold Company.

Waite, E. V. 1981. "Operational Results of National Solar Demonstration Projects." In L. Boyer, ed. *Earth Shelter Performance and Evaluation.* Stillwater, OK: Oklahoma State University Press.

Wells, M. and I. Spetgang. 1978. *How to Buy Solar Heating and Cooling . . . Without Getting Burnt.* Emmaus, PA: Rodale Press.

Wendt, R. L. 1982. *Earth Sheltered Housing: An Evaluation of Energy-Conservation Potential.* Oak Ridge, TN: U.S. Dept. of Energy, Oak Ridge National Laboratory.

Wright, D. and D. Andrejko. 1982. *Passive Solar Architecture.* New York: Van Nostrand Reinhold Company.

Chapter 3

A.I.A. Research Corporation. 1979. *A Survey of Passive Solar Buildings, HUD-PDR-287(2).* Washington, D.C.: U.S. Department of Housing and Urban Development.

Christensen, K. 1983a. "Send the Wind, Up, Around, and Away." *Earth Shelter Living,* vol. 27, 26–27.

Christensen, K. 1983b. "Use Vertical Fins and Barn Door Shutters." *Earth Shelter Living,* vol. 28, 9–11.

Givoni, B. 1976. *Man, Climate, and Architecture.* London, UK: Applied Science Publishers, Ltd.

Golany, G. 1983. *Earth Sheltered Habitat: History, Architecture, and Urban Design.* New York: Van Nostrand Reinhold Company.

Gray, D. and A. Leiser. 1982. *Biotechnical Slope Protection and Erosion Control.* New York: Van Nostrand Reinhold Company.

Jones, D. E., Jr. 1979. "The Expansive Soil Problem." *Underground Space,* vol. 3, no. 5, 221–226.

Kohler, J. and D. Lewis. 1981. "Passive Principles: Let the Sun Shine In." *Solar Age,* vol. 6, no. 11, 45–49.

Kubota, H. and N. Miley. 1981. "Thermal Analysis of a Passive Solar Earth Sheltered Home." In L. Boyer, ed. *Earth Shelter Performance and Evaluation Proceedings.* Stillwater, OK: Oklahoma State University Press.

Labs, K. 1980. "Earth Tempering as a Passive Design Strategy." In L. Boyer, ed., *Proceedings of Earth Sheltered Building Design Innovations.* Stillwater, OK: Oklahoma State University Press.

Labs, K. 1981a. *Regional Analysis of Ground and Aboveground Climate.* New Haven, CT: Undercurrent Design Research.

Labs, K. 1981b. "Regional Suitability of Earth Tempering." In L. Boyer, ed., *Earth Shelter Performance and Evaluation Proceedings.* Stillwater, OK: Oklahoma State University Press.

Labs, K. 1982a. "Regional Analysis of Ground and Above-Ground Climate Parts I and II." *Underground Space,* vol. 6, no. 6, 397–422.

Labs, K. 1982b. "Regional Analysis of Ground and Above-Ground Climate Conclusion." *Underground Space,* vol. 7, no. 1, 37–65.

Lynch, K. 1971. *Site Planning.* Cambridge, MA: M.I.T. Press.

Mattingly, G. and G. Peters. 1977. "Wind and Trees: Air Infiltration Effects on Energy in Housing." *Journal of Indoor Aerodynamics,* vol. 2, 1–19.

Mazria, E. 1979. *The Passive Solar Energy Book.* Emmaus, PA: Rodale Press.

Moffat, A. S. and M. Schiller. 1981. *Landscape Design That Saves Energy.* New York: William Morrow and Company.

Olgay, V. 1963. *Design With Climate.* Princeton, NJ: Princeton University Press.

Robinette, G. O. 1977. *Landscape Planning for Energy Conservation.* Reston, VA: Environmental Design Press.

Scalise, J. W. 1981. "A Survey of Earth Sheltered Housing in the Arizona-Sonoran Desert." In L. Boyer, ed., *Earth Shelter Protection.* Stillwater, OK: Oklahoma State University Press.

Schiller, M. 1982. "Landscape for Energy Efficiency." *Earth Shelter Living,* vol. 20, 45–47.

Underground Space Center, 1982. *Earth Sheltered Residential Design Manual.* New York: Van Nostrand Reinhold Company.

U.S. Department of Energy, 1981a. "Site Investigation." Earth Sheltered Structures Fact Sheet 1. Minneapolis, MN: Underground Space Center, University of Minnesota.

U.S. Department of Energy, 1981b. "Earth Coupled Cooling Techniques." Earth Sheltered Structures Fact Sheet 9. Stillwater, OK: School of Architecture, Oklahoma State University.

U.S. Department of Energy, 1981c. "Building in Expansive Clays." Earth Sheltered Fact Sheet 11. Stillwater, OK: School of Architecture, Oklahoma State University.

Chapter 4

Adams, J. 1982. "What to Look for in Window Insulation." *Solar Age,* vol. 7, no. 1, 46–55.

Anderson, B. 1983. *Underground Waterproofing.* Stillwater, MN: Webco Publishing.

Balcomb, D. 1981. "How to Balance Solar and Conservation in Passive Homes." *Solar Age,* vol. 6, no. 9, 38–45.

Balcomb, D. 1983a. "Balcomb's Final Guidelines." *Solar Age,* vol. 8, no. 11, 64.

Balcomb, D. 1983b. "Storing Heat in Concrete Masonry." *Earth Shelter Living,* vol. 26, 41–44.

Bargabus, D. 1982. "Computer Predicts Energy Costs." *Earth Shelter Living,* vol. 20, 55–56.

Becklian, B. 1980. "Homemade System Pipes in Savings." *Earth Shelter Living,* vol. 9, 6–7.

Blackford, J., and M. Curd. 1980. "Skylights: All the Beauty Without the Bugs." *New Shelter,* vol. 1, no. 3, 57–66.

Bregg, G. 1982. "Cost Overruns." *Earth Shelter Living,* vol. 23, 42–45.

Burton, J., and J. Reiss. 1980. "The Thermal Chimney." *New Shelter,* vol. 1, no. 5, 25–28.

Campbell, S. 1980. *Underground House Book.* Charlotte, VT: Garden Way Publishing.

Chamers, L. S., and J. A. Jones. 1980. *Homes In the Earth.* San Francisco, CA: Chronicle Books.

Earth Integrated Technics. 1983. *Earth Sheltered Plans for Better Living.* Stillwater, MN: Webco Publishing.

Elifrits, C., and A. Gillies. 1983a. "Earth Pipes: Preconditioning House Air." *Earth Shelter Living,* no. 29, 6–7.

Elifrits, C., and A. Gillies. 1983b. "Design of Air Tempering Facilities." *Earth Shelter Living,* no. 30, 26–27.

Francis, E. 1983. "Cooling with Earth Tubes." *Solar Age,* vol. 9, no. 1, 30–33.

Freyermuth, C. 1983. "The Post-Tensioning Industry, 1983—A Status Report." *Concrete Construction,* vol. 28, no. 4.

Givoni, B. 1976. *Man, Climate, and Architecture.* London, UK: Applied Science Publishers.

Gordon, A. 1980. "Plantings Provide Landscape Harmony." *Earth Shelter Living,* vol. 9, 10–11.

Hait, J. 1983. "Umbrella Modifies Soil Temperature." *Earth Shelter Living,* no. 27, 8–9. (See also *New Shelter,* vol. 4, no. 7, 98–100.)

Holthusen, T. L., ed. 1981. *Earth Sheltering: The Form of Energy and the Energy of Form.* New York: Pergamon Press.

Hylton, J. 1980. "Wind Towers Tested." *Earth Shelter Living,* vol. 11, 8–10.

Johnstone, P. 1982. "The Addition that Went Awry." *New Shelter,* vol. 3, no. 4, 34–35.

Kencil, D. 1980. "What's the Best Greenhouse Glazing?" *Earth Shelter Living,* vol. 10, 42–43.

Kimber, W. 1983. "Energy and Humidity Performance of "Total Wood" Earth Sheltered Homes." In L. Boyer, ed., *Earth Shelter Protection.* Stillwater, OK: Oklahoma State University Press.

Kis, B. 1980. "Grow Native Prairie Plants." *Earth Shelter Living,* vol. 9, 12.

Kliewer, T. 1982. *A Home for All Seasons.* Siloam Springs, AR: Tim Kliewer.

Kukula, K. 1983. "Smart Skylights." *New Shelter,* vol. 4, no. 9, 48–50.

Labs, K. 1982. "Living Up to Underground Design." *Solar Age,* vol. 7, no. 8, 34–38.

Lalo, J. 1980. "Cool Tubes." *New Shelter,* vol. 1, no. 5, 22–25.

Lane, C. 1982. "Underground Basics." *Popular Science,* vol. 221, no. 5, 76–79.

Langa, F. 1982a. "Cooling Without Kilowatts." *New Shelter,* vol. 3, no. 6, 17–22.

Langa, F. 1982b. "Insulate on the Outside." *New Shelter,* vol. 1, no. 3, 41–48.

Langley, J. B. 1981a. *Sun Belt Earth Sheltered Architecture—Part 1.* Winter Park, FL: John B. Langley.

Langley, J. B. 1981b. *Earth Sheltered Sun Belt Homes: 21 Floor Plans and Perspectives.* Winter Park, FL: John G. Langley.

Lowing, A. 1982a. "Earthquakes: Building in Seismic Zones." *Earth Shelter Living,* vol. 23, 13–14.

Lowing, A. 1982b. "Earthquake." *Earth Shelter Living,* vol. 24, 23–29.

Lowing, A. 1982c. "Earthquake: Architectural Detailing." *Earth Shelter Living,* vol. 25, 34–35.

Lunde, M. 1981a. "Efficiency Compared: Earth Covered and Thermal Roof." *Earth Shelter Living,* vol. 15, 26–29.

Lunde, M. 1981b. "Effects of Internal Mass." *Earth Shelter Living,* vol. 16, 23–25.

Lunde, M. 1981c. "Thermal and Covered Roof Costs Compared." *Earth Shelter Living,* vol. 18, 18–20.

Machowski, B. 1982. "Sunlight Piped Underground." *Earth Shelter Living,* vol. 21, 47–48.

Machowski, B. and K. Vadnais. 1982. "Builders Test Berm Spec Market." *Earth Shelter Living,* vol. 20, 14–18.

Mazria, E. 1979. *The Passive Solar Energy Book.* Emmaus, PA: Rodale Press.

McGroarty, B. 1980a. "Waterproofing: Sort Through the Myths." *Earth Shelter Living,* vol. 10, 10–11.

McGroarty, B. 1980b. "Waterproofing: Evaluation Backed By Experience." *Earth Shelter Living,* vol. 11, 23–26.

McGroarty, B. 1980c. "Waterproofing: Design to Work." *Earth Shelter Living,* vol. 12, 23–26.

McGroarty, B. 1981. "Waterproofing: Do Your Homework." *Earth Shelter Living,* vol. 13, 23–27.

Michels, T. 1979. *Solar Energy Utilization.* New York: Van Nostrand Reinhold Company.

Miller, T. 1982. "Aesthetics Should Be Considered." *Earth Shelter Living,* vol. 20, 58–59.

Mitchell, C. 1982. "One Man's Bevy of Heat Beaters." *New Shelter*, vol. 3, no. 6, 25–28.

Olgay, V. 1963. *Design With Climate*. Princeton, NJ: Princeton University Press.

Perkins, J. 1980. "Codes Slow Down Progress." *Earth Shelter Living*, vol. 10, 21–23.

Rawlings, R. 1982. "Double Shell Shakedown." *New Shelter*, vol. 3, no. 4, 28–29.

Roy, R. 1979. *Underground Homes*. New York: Sterling Publishing Company.

Rylander, R. 1980. "Vertical Crawl Space Developed." *Earth Shelter Living*, vol. 12, 8–9.

Shick, W. 1979. "Proper Building Orientation Can Save You Energy." *Earth Shelter Living*, vol. 2, 38–39.

Simmons, L. 1979. "Success With Residential Post-Tensioning." *Earth Shelter Living*, vol. 4, 23–27.

Simmons, L. 1981. "Response: Cover Your Roof." *Earth Shelter Living*, vol. 16, 16–19.

Skorusa, M. 1983. "Pull the Plug on Water Problems." *Earth Shelter Living*, vol. 28, 12–13.

Smolen, M. 1983a. "Natural Sidings." *New Shelter*, vol. 4, no. 3, 72–77.

Spears, J. 1982. "Goodbye Sloped Glass." *New Shelter*, vol. 3, no. 4, 30–34.

Stains, L. 1980. "The Sunspace: Building it for Yourself." *New Shelter*, vol. 1, no. 1, 37–44.

Szgethy, L. 1982. "Leaking Roof Fixed by Owner." *Earth Shelter Digest*, vol. 22, 40–42.

Tatum, R. 1978. *The Alternative House*. Danbury, NH: Addison House.

Traylor, E. 1981. "Build a Model Home as You Wait." *Earth Shelter Living*, vol. 14, 55–56.

Underground Space Center, University of Minnesota. 1978. *Earth Sheltered Housing Design: Guidelines, Examples, and References*. New York: Van Nostrand Reinhold Company.

Underground Space Center, University of Minnesota. 1981. *Earth Sheltered Homes: Plans and Designs*. New York: Van Nostrand Reinhold Company.

Underground Space Center, University of Minnesota. 1982. *Earth Sheltered Residential Design Manual*. New York: Van Nostrand Reinhold Company.

U.S. Department of Energy. 1981a. "Daylighting Design." Earth Sheltered Fact Sheet 7. Stillwater, OK: School of Architecture, Oklahoma State University.

U.S. Department of Energy. 1981b. "Indoor Air Quality." Earth Sheltered Fact Sheet 8. Stillwater, OK: School of Architecture, Oklahoma State University.

U.S. Department of Energy. 1981c. "Passive Solar Heating." Earth Sheltered Fact Sheet 12. Stillwater, OK: School of Architecture, Oklahoma State University.

Vadnais, K. 1979. "Rain Plagues Builder." *Earth Shelter Living*, vol. 2, 25–27.

Vadnais, K. 1980. "Covered vs Conventional: A Friendly Debate." *Earth Shelter Living*, vol. 11, 27–33.

Vadnais, K. 1981. "Old Arch, New Material Combined." *Earth Shelter Living*, vol. 17, 6–11.

Vadnais, K. 1982a. "Drainage System is Different." *Earth Shelter Living*, vol. 22, 19.

Vadnais, K. 1982b. "Steel House Goes On Market." *Earth Shelter Living*, vol. 23, 34–37.

Vadnais, K. 1982c. "Warranty Backs Earth Shelter." *Earth Shelter Living*, vol. 19, 12–13.

Vadnais, K. 1983a. "Drainage: Options Are Growing." *Earth Shelter Living*, vol. 27, 34–35.

Vadnais, K. 1983b. "Insulation: Puddle Rebuttal." *Earth Shelter Living*, vol. 28, 4–5.

Viceps, K. 1982. "Plan for Leaks." *Earth Shelter Living*, vol. 23, 16–17.

Villagran, N. 1982. "Domes Form Room Clusters." *Earth Shelter Living*, vol. 21, 20–22.

Wade, H. 1980. "Earth Sheltering's Newest Technique." *New Shelter*, vol. 1, no. 7, 44–48.

Wade, H. 1983. *Building Underground*. Emmaus, PA: Rodale Press.

Wells, M. 1977. *Underground Designs*. Brewster, MA: Malcolm Wells.

Wells, M., and S. Glenn-Wells. 1980. *Underground Plans Book-1*. Brewster, MA: Malcolm Wells.

Williams, M. 1980. "Heat Exchanger Uses Water Efficiently." *Earth Shelter Living*, vol. 10, 40–41.

Woodrum, D. 1982. "Earth Cover Expense is Balanced." *Earth Shelter Living*, vol. 24, 10–17.

Woodrum, D. 1983. "Underfloor Pipes Control Passive Gain." *Earth Shelter Living*, vol. 29, 14–15.

Wright, D. 1978. *Natural Solar Architecture: A Passive Primer*. New York: Van Nostrand Reinhold Company.

Chapter 5

Anderson, B. 1983. *Underground Waterproofing.* Stillwater, MN: Webco Publishing.

American Concrete Institute. 1978. *ACI Manual of Concrete Practice: Recommended Practice for Measuring, Mixing, Transporting, and Placing Concrete, No. ACI 304-73.* Detroit, MI: American Concrete Institute.

Bureau of Naval Personnel. 1972. *Basic Construction Techniques for Houses and Small Buildings.* New York: Dover Publications. American Concrete Institute.

Campbell, S. 1980. *The Underground House Book.* Charlotte, VT: Garden Way Publishing.

Carter, D. 1982. *Build it Underground.* New York: Sterling Publishing Company.

Concrete Construction Magazine, 1980. *Earth Sheltered Construction.* Addison IL: Concrete Construction Publications, Inc.

Dick, C. 1981. "Add Mixture, Not Water." *Earth Shelter Living,* vol. 17, 23.

Goldberg, L. 1983. "Underground Space Center: Monitoring Program." *Earth Shelter Living,* vol. 29, 33–35.

Gray, D. and A. Leiser. 1982. *Biotechnical Slope Protection and Erosion Control.* New York: Van Nostrand Reinhold Company.

Hait, J. 1983. "Umbrella Modifies Soil Temperature." *Earth Shelter Living,* vol. 27, 8–9.

High Pressure Shotcreting Corporation. 1982. "Concrete Can Be Sprayed." *Earth Shelter Living,* vol. 20, 20–22.

Holland, E. 1981. "Insulation Moves Outside." *Solar Age,* vol. 6, no. 11, 22–27.

Kern, K., B. Kern, J. Mullan, and O. Mullan. 1982. *The Earth Sheltered Owner-Built Home.* North Fork, CA: Owner-Builder Publications.

Kimber, W. 1983. "Energy and Humidity Performance of "Total Wood" Earth Sheltered Homes." In L. Boyer, ed. *Earth Shelter Protection.* Stillwater, OK: Oklahoma State University Press.

Langley, J. 1980. "The Barrel Shell—Structural Rethinking in Earth Sheltered Design" *Underground Space,* vol. 5, no. 2, 92–101.

Langley, J. 1981. *Sun Belt Earth Sheltered Architecture.* Winter Park, FL: John B. Langley.

McGroarty, B. 1980a. "Waterproofing: Sort Through Myths." *Earth Shelter Living,* vol. 10, 10–11.

McGroarty, B. 1980b. "Waterproofing: Evaluation Backed by Experience." *Earth Shelter Living,* vol. 11, 23–24.

McGroarty, B. 1980c. "Waterproofing: Design to Work." *Earth Shelter Living,* vol. 12, 23–25.

McGroarty, B. 1981. "Waterproofing: Do Your Homework." *AU (Earth Shelter Living),* vol. 13, 23–27.

Meixel, G., P. Shipp, and T. Bligh. 1980. "The Impact of Insulation Placement on the Seasonal Heat Loss Through Basement and Earth Sheltered Walls." *Underground Space,* vol. 5, no. 1, 41–47.

Ropke, J. 1982. *Concrete Problems, Causes and Cures.* New York: McGraw-Hill Book Company.

Roy, R. 1979. *Underground Houses.* New York: Sterling Publishing Company.

Scott, R. 1979. *How To Build Your Own Underground Home.* Blue Ridge Summit, PA: Tab Books.

Simmons, L. 1979. "Success With Residential Post-Tensioning." *Earth Shelter Living,* vol. 4, 23–27.

Slater, D. 1982. "Backfill Properly to Prevent Damage." *Earth Shelter Living,* vol. 22, 50–51.

Sterling, R. 1978. "Structural Systems for Earth Sheltered Housing." *Underground Space,* vol. 3, no. 2, 75–81.

Sterling, R. and M. Tingerthal. 1981. "Building Costs and Construction Problems in the Minnesota Earth-Sheltered Housing Demonstration Program." *Underground Space,* vol. 6, no. 1, 13–20.

Szigethy, L. 1982. "Leaking Roof Fixed by Owner." *Earth Shelter Living,* vol. 22, 40–42.

Underground Space Center, University of Minnesota. 1982. *Earth Sheltered Residential Design Manual.* New York: Van Nostrand Reinhold Company.

U.S. Department of Energy. 1981a. "Insulation Principles." Earth Sheltered Structures Fact Sheet, No. 5. Underground Space Center, University of Minnesota.

U.S. Department of Energy. 1981b. "Insulation Materials and Placement." Earth Sheltered Structures Fact Sheet, No. 6. Underground Space Center, University of Minnesota.

Vadnais, K. 1982. "House Full of Unusual Characteristics." *Earth Shelter Living,* vol. 22, 26–29.

Wade, H. 1983. *Building Underground.* Emmaus, PA: Rodale Press.

Chapter 6

Boyer, L., W. Grondzik, and M. Weber. 1980. "Passive Energy Design and Habitability Aspects of Earth-Sheltered Housing in Oklahoma." *Underground Space*, vol. 4, no. 6, 333–339.

Boyer, L. and W. Grondzik. 1983a. "Comfort Assessment in Earth Covered Dwellings in the United States." In L. Boyer, ed. *Earth Shelter Protection.* Stillwater, OK: Oklahoma State University Press.

Boyer, L. and W. Grondzik. 1983b. "Energy Performance of Earth Covered Dwellings in the U.S." In L. Boyer, ed. *Earth Shelter Protection.* Stillwater, OK: Oklahoma State University Press.

Bruno, R. 1983. "Sources of Indoor Radon in Houses: A Review." *Journal of the Air Pollution Control Association*, vol. 33, no. 2, 105–109.

Brzezowski, E. and R. Kirchner. 1981. "Powerful Monitor is Simple, Inexpensive." *Solar Age*, vol. 6, no. 8, 55–56.

Feisel, L. 1980. "Monitoring System Defined." *Earth Shelter Living*, vol. 9, 34–35.

Fuller, W. 1981. "What's in the Air for Tightly Built Houses?" *Solar Age*, vol. 6, no. 6, 30–32.

Goldberg, L. 1983. "Underground Space Center: Monitoring Program." *Earth Shelter Living*, vol. 29, 33–35.

Holon, S., P. Kendall, S. Norsted, and D. Watson. 1980. "Psychological Responses to Earth-sheltered, Multilevel, and Aboveground Structures With and Without Windows." *Underground Space*, vol. 5, no. 3, 171–178.

Landa, E. 1983. "Radon Concentrations in the Indoor Air of Earth Sheltered Buildings in Colorado." In L. Boyer, ed. *Earth Shelter Protection.* Stillwater, OK: Oklahoma State University Press.

Lord, D. 1981. "Interior Environmental Quality in Earth Shelters." In L. Boyer, ed. *Earth Shelter Performance and Evaluation.* Stillwater, OK: Oklahoma State University Press.

Machowski, B. 1982a. "Oldest Fuel is Updated." *Earth Shelter Living*, vol. 23, 10–11.

Machowski, B. 1982b. "Solar Tax Credits." *Earth Shelter Living*, vol. 7, 33–41.

May, H. 1981. "Ionizing Radiation Levels in Energy-Conserving Structures." *Underground Space*, vol. 5, no. 6, 384–391.

McKown, C. and K. Stewart. 1980. "Consumer Attitudes Concerning Features of an Earth-Sheltered Dwelling." *Underground Space*, vol. 4, no. 5, 293–295.

Nero, A. 1983. "Radon in Energy-Efficient Earth Sheltered Structures." In L. Boyer, ed. *Earth Shelter Protection.* Stillwater, OK: Oklahoma State University Press.

Oswald, R. 1983. "Tight House Can Seal in Pollution." *Earth Shelter Living*, vol. 27, 36–37.

Paul, T. 1982. "How to Design a Remote Power System." *Solar Age*, vol. 7, no. 10, 34–39.

Pick, E. 1980. "Operations Characteristics of a Utility-Free Dwelling in Kansas." In L. Boyer, ed. *Earth Sheltered Building Design Innovations.* Stillwater, OK: Oklahoma State University Press.

Rand, G. 1981. "Think Ecologically About Indoor Environments." *Underground Space*, vol. 6, no. 2, 105–108.

Rollwagen, M., S. Taylor, and T. Holthusen. 1983. "Buying an Existing Earth-Sheltered Home." *Mother Earth News*, vol. 83, 84–85.

Seitz, D. 1983. "Reflecting on Photovoltaics." *Earth Shelter Living*, vol. 30, 10–12.

Selinfreund, M., R. Farrer, and P. Munding. 1983. "Monitoring an Earth-Sheltered Solar-Assisted House." In L. Boyer, ed. *Earth Shelter Protection.* Stillwater, OK: Oklahoma State University Press.

Shurcliff, W. 1980. *Thermal Shutters and Shades.* Andover, MA: Brick House.

Shurcliff, W. 1983. *Air-to-Air Heat Exchangers for Houses.* Andover, MA: Brick House.

Stickney, B. 1978. "The Homeowner's System for Evaluating Passive Solar Heating." *Passive Systems '78.* Newark, DE: International Solar Energy Society.

Vadnais, K. 1980. "Light of Financial Breaks Shines on Passive Solar." *Earth Shelter Living*, vol. 7, 28–31.

Vadnais, K. 1983. "Insurance Study Favors Earth Sheltering." *Earth Shelter Living*, vol. 26, 26–28.

Wadden, R. and P. Scheff. 1983. *Indoor Air Pollution, Characterization, Prediction, and Control.* New York: John Wiley and Sons.

Appendix A

Balcomb, J. D.; Barley, D.; McFarland, R.; Perry, J.; Wray, W.; and Noll, S. 1980. *Passive Solar Design*

Handbook, Vol. II, p. 15. United States Department of Energy, DOE CS-0127/2.

Bligh, T. 1976. Energy conservation by building underground. *Underground Space* 1 (1): 19–23.

Blick, E. F. 1980. A simple method for determining heat flow through earth covered roofs. *In Proceedings of Earth Sheltered Building Design Innovations Conference,* ed., L. L. Boyer. Stillwater, OK: Oklahoma State University.

Boyer, L. L. 1980. Energy usage in earth covered dwellings in Oklahoma. In *Proceedings of Earth Sheltered Building Design Conference,* ed., L. L. Boyer. Stillwater, OK: Oklahoma State University.

Brown, G. Z. and Novitski, B. 1981. Climate responsive earth sheltered buildings. *Underground Space* 5 (5): 229–305.

Campbell, S. 1980. *The Underground House Book,* p. 194. Charlotte, VT: Garden Way.

Green, K. W. 1979. Passive cooling: designing natural solutions to summer cooling loads. *Research and Design,* AIA Research Corporation 11 (3): 4.

Labs, K. 1980. Earth tempering as a passive design strategy. In *Proceedings of Earth Sheltered Building Design Innovations Conference,* ed., L. L. Boyer. Stillwater, OK: Oklahoma State University.

Mazria, E. 1979. *The Passive Solar Energy Book,* 166–167. Emmaus, PA: Rodale Press.

Morrison, J. 1979. *The Kansas Energy Saving Handbook for Homeowners,* p. 213. New York: Harper and Row.

Simmons, L. B. 1979. Success with residential post-tensioning. *Earth Shelter Digest* 1 (4): 23–27.

Smith, D. L. 1979. Mean radiant temperature and its effects on energy conservation. In *Proceedings of the Fourth National Passive Solar Conference,* ed., G. Franta. International Solar Society.

Sterling, R., ed. 1978. *Earth Sheltered Housing Design,* p. 51. Underground Space Association, University of Minnesota.

Szydlowski, R. and Kuehn, T. 1980. Transient analysis of heat flow in earth sheltered structures. In *Proceedings of Earth Sheltered Building Design Innovations Conference,* ed., L. L. Boyer. Stillwater, OK: Oklahoma State University.

INDEX

NOTES

NOTES